W9-AGB-997

AMERICAN AUTOBAHN

The Road To An Interstate Freeway With No Speed Limit

Mark Rask

with illustrations by

Mark Stehrenberger

Vanguard Non-Fiction Books
Minneapolis • Minnesota

Published by:
Vanguard Non-Fiction Books
P.O. Box 22370
Minneapolis, Minnesota
55422-0370

Visit our website at:
www.americanautobahn.com

Printed in Canada

10 9 8 7 6 5 4 3 2 1
First Edition

ISBN: 0-9669136-0-4

LCCN: 98-90867

Cover photo by the author
Back cover photo by Mike Valente

To
Louis C. Wendling,
a driver's driver,
and a true automotive enthusiast

CONTENTS

Every start upon an untrodden path is a venture which only in unusual circumstances looks sensible and likely to be successful.

Albert Schweitzer

INTRODUCTION

Not Just a Faster, Safer Road but a Better America as Well

I was a living, breathing part of the American Dream. . .

The late October sun hung low and cool in the morning sky as I drove my father's navy blue, 1965 Chrysler Newport near the Wyoming-Montana border. On an empty stretch of U.S. Highway 212, the squared-off, four-door sedan plowed through the air with all the aerodynamics of a sledgehammer. The 383 V-8 droned out the power to push the speedometer needle to the 80 mph mark—five over Wyoming's posted limit.

The big Chrysler floated effortlessly down the rolling two-lane highway, the occasional bump smoothed out by the wallowing, mattresslike ride that drifted slowly to the left or right as I, one-handed, twisted my wrist to bring the errant car back straight. My other hand worked to push an uncooperative safety belt into the recesses of the sofalike seat.

As we approached the border, my conversation with a high-school buddy of mine hung on the indignation of having to pay thirty-six cents for a gallon of gas *and* having to pump it yourself. The 1973 Arab oil embargo was in full swing. Just a few weeks earlier we had paid a whopping twenty-four-nine for full-service regular.

Shooting past the Montana border sign, we sped by another, larger one:

SPEED LIMITS

DAY — REASONABLE & PRUDENT

NIGHT — 55

TRUCK — 50

I wondered what in the world "reasonable and prudent" could mean exactly. After a few moments' debate, we agreed it had to be more than 80 mph. Soon, the speedometer needle wavered either side of 105. The high-pitched whistle of the wind whipped past the side mirrors as the balding, bias-ply snow tires whined over the dull moan of the engine. All was a blur to the left and right, but the road ahead and the foothills beyond were crystal clear.

Suddenly, the rearview mirror was filled with the chrome grille and white hood of a Dodge sedan. I glanced further up the mirror to catch the roofline and froze: The light bar and siren rack of the Montana Highway Patrol—lights not yet flashing.

I didn't brake, didn't move. I was caught dead to rights at 105 mph. Out of the corner of my mouth, I told my friend not to look back at what was traveling in tandem just a car length behind us.

After a few nervous seconds, the cruiser fell back slightly, pulled out and up alongside of us. I looked over sheepishly as the trooper shot us a quick smile and salute, kicked down the 440 wedge engine of the big sedan and left us in the dust. I looked over to my buddy with a dropped jaw that turned to a smile of glee.

We continued on at our reasonable and prudent speed of 105 mph.

This youthful tale of two seventeen-year-old boys illustrates the many contradictions associated with driving in America. That carefree, happy abandon that comes from an immature love affair—blithely driving down the road without recognizing the potentially deadly consequences. In the past quarter century, this American love affair with driving has waned only slightly and matured little. Shaken by two oil embargoes, an unrealistic and arbitrary national maximum speed limit, a maze of federal regulation, deregulation, gridlock, crumbling infrastructure and environmental concerns, it's a miracle this love affair endures at all! Yet, it *does* endure and continues to be a unique part of the American experience.

There is no danger of this love affair becoming anything less than what it already is, but if it were to move beyond infatuation to a more mature and responsible relationship with the automobile and driving, this bond would be even stronger. The biggest hindrance to this future enjoyment of the car and the open road is the continued death and suffering it brings to us each year. Since 1896, more than 3 million people have died on America's roads, with countless more injured and disabled. When compared to other forms of travel such as air, boat or rail, 94 percent of all transportation fatalities are caused by motor vehicles. The reason we haven't been able to reduce these numbers significantly is because we fail to recognize the biggest single killer on the road: our attitude towards driving as it relates to safety and speed.

Historically, we have openly condemned "speeding" as a dangerous, guaranteed, one-way ticket to the graveyard, spending hundreds of millions of dollars every year trying to hold highway speeds down. For twenty-one years, traffic was forced to crawl, in the name of safety, along half a million miles of rural, two- and four-lane highways posted at 55 mph. The same was true of our interstate freeway system, but mounting pressure from the driving public forced the federal government in 1987 to allow states to raise that limit to 65 mph on rural stretches. Finally, in December 1995, federal speed-limit control was repealed altogether as motorists voted not with the ballot box, but the gas pedal. Still, the official anthem of the highway safety movement in this country remains, "Speed Kills." To be in favor of higher speeds and faster travel is to invite more death and suffering on the road.

The driving public, however, continues to ignore this safety battle cry. With a wink-and-a-nudge and a don't-get-caught attitude, America, as a nation, is now driving faster than ever before. State after state gives into this trend by raising speed limits on more and more roads. Automobile manufacturers goad customers with suggestive advertising that hypes the thrill of the open road and the power on tap when you put your foot down, pedal-to-the-metal. Radar-detector companies, first discreetly, then with respectability, openly sell millions of high-tech "fuzzbusters" for evading police radar. The cops fight back with higher-tech laser speed-sensing devices. Motorists up the ante, arming themselves with the latest laser detectors and jammers. Still, the federal government estimates that law enforcement officers write well over 10 million speeding tickets every year. More importantly, the percentage of motorists exceeding the speed limit on most roadways ranges from 50 to 95 percent.

Clearly, the fact of the matter is, America wants to drive *faster*, not slower. Our two-decade-long attempt to force America to slow down was a dismal failure. In the process, we became a nation fighting ourselves. The consequence of this preoccupation with speed enforcement and our thirst for speed was—and is—our inability to lower dramatically the number of people killed and injured on the road. This is not to say that repealing 55 and stepping on the gas has improved safety, either. Driving faster entails some real responsibility and innovative action to be done safely. If successfully combined, safety and speed would provide a boon to our economy. The American auto industry would surge forward with the latest innovations, building superior vehicles to rival the foreign competition. Fewer fatalities would lower medical and auto insurance premiums as productivity and transportation efficiency increase. Interstate commerce would move quicker and cheaper than ever before, unfettered by excessive fines, fees and regulations. With the highest standards of safety, motorists could drive 300 miles in under three hours. Our outlook would improve as well, working *together* to solve a serious national problem.

Such an accomplishment is not some mythical impossibility. This book takes half its name from the marvelous system of freeways that crisscross the country of Germany. Even though most rural stretches of this 7,000-mile freeway network have *no* speed limit, the death rate for the *Autobahn* is *lower* than the rate for our interstates posted at 55, 65 or 75 mph. This has happened in spite of the fact that traffic on the *Autobahn* routinely flows above 95 mph. As a result of this demanding driving environment, German automobiles set an unrivaled standard of safety at high speed, for which they are sought after and exported all over the world. But such achievements are far more complex than simple statistics suggest. They don't tell the whole story. So what follows is a straightforward, honest appraisal of why Germany has succeeded in saving lives, where America has gone wrong and how we could improve both safety and speed on the open road.

From the automotive enthusiast behind the wheel of a fast car, to the slow driver who gets an angry knot in the stomach every time the word speed is mentioned, what follows may surprise you.

CHAPTER 1

Glory Days 1940–1973

Day after day, rain or shine, the people lined up and waited for hours outside the General Motors "Highways and Horizons" exhibit at the 1939 New York World's Fair. Sometimes standing 15,000 deep, these patient souls hoped to become one of the lucky 600 spectators to view the Futurama of 1960. Once inside, each succeeding group queued up to ride a train of chairs through a miniature model of the America of tomorrow. A synchronized recording in each chair described the passing wonders as the passengers marveled at the gleaming, sun-drenched cities alongside rolling prairies and mountains. The thread binding this coast-to-coast utopia was a network of superhighways, some fourteen lanes wide, with teardrop-shaped cars and trucks whizzing down computer-controlled lanes at speeds of 50, 75 and 100 mph. The typical intersection of the day had become a multilaned overpass of swirling cloverleafs and four-pointed diamonds. The expressway itself bypassed smaller towns and potential traffic jams, taking advantage of all the automobile had to offer. The exhibit expressed the viewpoint of General Motors and the designer of the Futurama, Norman Bel Geddes, that "Automobiles are in no way responsible for our traffic problem. . . . The entire responsibility lies in the faulty roads." To an America still driving on narrow, winding, two-lane

The Road to Tomorrow. The General Motors Futurama Exhibit at the 1939 New York World's Fair. ©General Motors Corp. Used with permission of GM Media Archives

highways, this sixteen-minute show was a dream come true, and the Futurama quickly became the Fair's most popular attraction. Other fairgoers could easily spot the lucky ones who had already experienced the exhibit, each sporting a small, blue-and-white lapel pin declaring, "I have seen the future." Demand was so great to see the Futurama that it remained open through 1940 with a total of sixteen million people visiting GM's automotive view of tomorrow.

As the last of these visitors lined up to ride through the General Motors exhibit, the finishing touches were being put on America's first limited-access toll road 150 miles to the west: The Pennsylvania Turnpike. Carved out of the Allegheny Mountains along the abandoned right-of-way for the unfinished South Penn Railroad, this $70 million, 160-mile-long superhighway was the closest thing yet to the expressways of GM's Futurama. Nothing like it had been built in this country before. Stretching from just outside the state capital of Harrisburg to the outskirts of Pittsburgh, the road cut its way through some of the most rugged terrain in the state. Yet, the grade was no greater than a three-foot rise or fall for every 100 feet traveled (3 percent grade), guaranteed by seven tunnels bored straight through the mountains. The neighboring Lincoln Highway was a two-lane thoroughfare with 9 percent grades, 939 cross streets, twelve railroad crossings and twenty-five traffic signals over the same distance. The new turnpike eliminated every at-grade intersection and stoplight, in line with the new philosophy of making the road conform to the automobile instead of vice versa. A 200-foot right-of-way was cut through the countryside, starting with ten-foot-wide shoulders, four twelve-foot-wide concrete lanes and a ten-foot median strip to separate the cross traffic. Superelevation (banking) was used on all curves and each of the eleven tollbooth access points had 1200-foot acceleration/deceleration lanes for vehicles entering or leaving the turnpike. With three-quarters of the road's length free of curves, and cut into the south and west side of hills for the sun to melt ice and snow, safety officials were convinced this all-weather highway would eliminate 95 percent of all causes of accidents.

With such safety innovations a part of its design, the turnpike was set to open in the Fall of 1940 without a speed limit clearly established by law. This was not all that uncommon at the time. Several states did not have prima-facie speed limits, but relied on the vague "reasonable and prudent" for their basic speed law on rural two-lane highways. However, the commonwealth of Pennsylvania did have a statewide 50 mph speed limit, and two weeks before the opening of the turnpike, Governor Arthur H. James declared that this limit would apply to the new toll road. This was decided in spite of the fact that a test car had hit a speed of 102 mph on the new superhighway and traffic engineers reassured everyone this kind of speed was "safe." (The average top speed for most cars of the day was between 85 and 90 mph.)

Given only twelve hours advance notice before the grand opening, motorists quickly lined up at the eleven tollbooths to be the first to drive the new turnpike. A few minutes past midnight on October 1, one toll attendant put his hand down race-car style and hundreds of cars and trucks were off and running. With an almost circuslike atmosphere, virtually everyone was streaking down the new road at breakneck speeds. Many drivers covered the 160-mile distance in

"America's Dream Road," The Pennsylvania Turnpike opened for travel on October 1, 1940—with no speed limit clearly established by law. Pennsylvania Turnpike Commission

record time, averaging 75 mph in spite of seven 35 mph two-lane tunnels and the steady stream of traffic. More than one driver promptly pulled over at the sight of the light gray squad cars of the Pennsylvania Motor Police *and* a speedometer reading of 70 to 80 mph. The troopers responded with lighthearted small talk about the new transportation marvel and no speeding tickets, worrying more about courtesy on the road and staying in your own lane. The new turnpike dramatically cut travel time between Harrisburg and Pittsburgh from six hours to, in many cases, under two-and-a-half hours. One trucker marveled that he saved six-hours time and twenty gallons of gas. By day's end, those who personally drove on the new superhighway realized there was no speed limit for this first-ever freeway in America.

Officially, however, the press played down the high speed of the event. The Harrisburg *Patriot* declared on October 2:

> "A speed of fifty miles an hour has been designated by Governor James but one motorist averaged nearly ninety miles an hour over the road yesterday. This was noted by a toll house attendant who said the motorist's ticket had been stamped 8:35 a.m.

> at Bedford. He arrived at the Carlisle Interchange at 9:27 a.m. [79 miles]. No speed limit signs have been erected on the new highway as yet. Attendants when asked about the speed by motorists, merely reply "drive carefully."

In the days that followed, the paper reiterated the governor's edict but cryptically informed the public that "many car drivers have exceeded that speed without danger of arrest by the Motor Police who patrol the new road." The governor himself took to the turnpike on October 8, declaring it a "peach of a road," and "the most somnolent ride I ever had," taking four hours to cover the 160-mile distance at just under 50 mph. He advised that everyone should be given a cup of coffee to keep them awake at such a speed. In the lore of the turnpike, some say, rumors persisted that at least part of his trip was driven much faster, and this only added to the contradiction of what was officially talked about and what was really taking place.

As the weeks passed, scores of people came from all over the country to drive on the new expressway. Huge traffic jams ensued, especially on the weekends. On October 20, a man from Bethlehem, Pennsylvania skidded off the road, struck a bridge and became the first person to die on the turnpike. By year's end, four more people were killed on the road as a grand total of 640,000 cars and trucks safely negotiated the high-speed freeway in the first three months it had been open. Through the winter, the "official" speed limit was maintained in the press, but the only actual restrictions for drivers on the toll road were weather, the flow of traffic and their cars. Many vehicles suffered from overheated engines and blown tires as drivers tried to find out just how fast their cars would go. Advertising also began to influence the public's attitude toward the high-speed turnpike. One ad from Ford Motor Company stated, "The closest the average American comes to breaching the sonic barrier is when he eases himself behind the wheel of the family car and has a go at the Pennsylvania Turnpike."

By spring, the official stance was a public joke. It had become plainly evident that no speed limit was in force and pressure within the state government was growing to do something about it. Turnpike engineers maintained that speeds of 100 mph could be safely driven on the road, but in April 1941 a compromise was reached. On the 15th, the governor signed into law a maximum turnpike speed limit of 70 mph. Governor James was so worried that he would have blood on his hands, he signed a disclaimer along with the bill, turning authority over to the Turnpike Commission "not to permit speeds which prove to be excessive." From that day forward, there was no doubt what the posted limit of the turnpike would be. But what the law stated and what was really taking place out on the open road were moving further apart at an ever greater rate of speed.

After the Japanese bombed Pearl Harbor on December 7, 1941, World War II brought radical change to America. The entire country mobilized into a unified force with a speed and organization unparalleled in history. Sacrifice, hardship and cooperation became the watchwords of the time. Regulation and rationing were now dictated from the federal level. In an attempt to reduce gasoline use, save tire rubber and instill a spirit of national resolve, a National Maximum Speed Limit of 35 mph, the first of its kind, was imposed from coast-to-coast. Average speeds for passenger cars dropped from 48 mph to 37 mph as gas rationing and military mobilization brought all but necessary travel to a standstill. The number of miles driven fell off sharply, and the number of people killed on the road in 1941 and '42 plunged from 38,142 to 27,007. These facts went all but unnoticed as public attention focused on the war effort. With thousands of Americans being shipped overseas, such an exodus shifted some casualties from the road to the battlefield. Though difficult ever to know for sure, for a nation still driving on narrow, two-lane highways, the slower speeds probably played a part in lowering the death toll and minimizing injuries. As the war swung in our favor, more drivers began exceeding the "Victory Speed Limit" of 35 mph. By the time the limit was repealed in August 1945, it was back to business as usual. The influx of military men and women moving back into civilian life created a boom in the amount of road traffic, and the number of people killed in 1946 jumped to 31,874. However, the death rate (based on the number of people killed per 100 million vehicle-miles traveled) continued on its historic decline.

You Have to Slow Down to 90 for the Curves

A typical example of Goodrich improvement in tires

IMAGINE a flat, double ribbon of highway stretched taut over valleys, slicing between hills, running straight through huge tunnels bored in mountains. The grades so slight you hardly know you're climbing. Not a single stoplight or a crossroad in 160 miles. No speed limit. Few curves—and all of those can be taken at 90 miles per hour.

It's a dream highway—a dream that has come true. It's the new Pennsylvania Turnpike between Pittsburgh and Harrisburg. Building this super-express highway was a herculean project—a fantastic story of engineering triumph. Considered a three or four-year job, it was completed in approximately 20 months, employing as many as 25,000 men and thousands of trucks at one time. While one crew cut down a mountain, others tore tunnels through solid rock, still others worked feverishly erecting the 285 necessary bridges.

Rubber made this highway possible! Special Goodrich Earth Mover and Rock Service Tires developed by engineers of The B. F. Goodrich Company were used on hundreds of trucks and earth-moving vehicles. Some of this equipment was so large that it picked up 25 to 30 tons of earth and rock, placed it in 8-inch layers, and rolled it—all in as short a time as 12 minutes. Without these new man-high tires making possible gigantic loads, the Pennsylvania Turnpike probably would not have been built!

The development of these special tires is typical of the research work done by Goodrich to give you better truck and bus tires. For every transportation problem Goodrich provides special tires designed to give the longest possible wear for that type of service. If you want to cut your tire costs, call the Goodrich man. Remember which, the name's Good*rich*. The B. F. Goodrich Co., Akron, O.; Los Angeles, Calif.; Kitchener, Ontario.

Goodrich Silvertowns
FOR TRUCKS AND BUSES

PHILADELPHIA	HARRISBURG	PITTSBURGH
1955 Hunting Park Ave.	2nd and North Sts.	5740 Baum Blvd.

SEPTEMBER, 1940 *Patronize HIGHWAY BUILDER Advertisers*

Courtesy of B.F. Goodrich Company

Curiously, speeds remained relatively flat, cruising along at an average of 48 mph for several years afterward.

For the most part, these facts and figures were noted only by statisticians as America's coming home from war brought with it a pent-up desire to enjoy and celebrate. After ten years of depression and five years of war, the country had earned a long overdue vacation from sacrifice. This created an ever growing need and want to buy things: homes, appliances and, especially, *cars*. With growing demand and little preparation (no passenger cars were manufactured from 1942 to 1945), the 1946 models were little more than cobbled-up versions of 1941 automobiles. Demand was so great, many new-car buyers were putting money down under the table only to be out-bribed by someone else. It did not last long, however. Detroit met the challenge, and soon new models with snappier styling appeared on the market, being quickly bought up by a car-hungry America.

By 1948, and the introduction of the automatic transmission by General Motors, the standard American passenger car had achieved a certain mechanical plateau. Manufacturing automobiles with front independent suspension, hydraulic drum brakes, dependable engine and transmission, the "Big Three," Ford, Chrysler, and General Motors, were beginning to sit back and rest comfortably with a chassis that would permit a long and profitable production of cars without any glitches.

One maverick entering the automotive scene was convinced something basic and fundamental was lacking. Industrialist Preston Tucker, after a certain amount of business and political anguish, introduced a state-of-the-art prototype car that put Detroit and the rest of America on its ear. Sculpted out of a fresh, clean design, the Tucker '48 sported a powerful six-cylinder engine mounted in the *rear* of the car, four-wheel disc brakes—unheard of at the time—and a center-mounted headlight which turned with the front wheels, illuminating the road around corners. These innovations alone would have been enough to cause concern in Detroit, but the interior of Tucker's car was even more controversial. It had been designed to protect passengers if they were involved in a collision. Up until this time, shatterproof safety glass had been the only protection given passengers in a crash. The prevailing attitude among law enforcement and the driving public was that an accident was such a horrendous event, there was no way to survive it except to avoid it in the first place. In the Tucker '48, padding was placed on the seat backs and dashboard, door handles were flush mounted and the rearview mirror would break away when hit. Control knobs were rounded and placed out of the way on the left side of the steering column. The glove box was replaced with a "safety chamber" that the driver and front passenger could dive into just before an accident. Even the windshield would pop out when one hundred pounds of force was applied from the inside. Future models promised such unusual innovations as a padded, collapsible steering wheel and speedometer on the hood so the driver would not have to look down from the road. The country seemed mesmerized and was soon caught up in Tuckermania. Dealership options were snapped up as prospective customers put money down. The combination of style, performance and "safety" interior had captured the public's fascination. The promise of futuristic technology and the manufacturer's attitude towards safety seemed to all but ensure success for Tucker's fledgling car company.

This new approach to interior safety didn't happen by accident. It was due to the direct influence of a plastic surgeon from Detroit. Dr. Claire L. Straith had been trying for years with little success to influence the other car manufacturers to make their interiors more "friendly" to passengers. Having seen first hand what sharp edges and spearlike switches could do to a person's body and face, Straith first modified his own car in the 1930s to see if he could improve on the problem. Installing seatbelts and padding around the dash, he was inspired to design several more refined dashboards using springs and padding to absorb the shock of a person's legs, face or body striking it in a collision. After patenting these designs in 1935 and '37, he set out to influence the carmakers' interior designs, having mild success with Chrysler and the 1937 Dodge. The interior sported flush instrument panel buttons when not in use, a raised dashboard to minimize the risk of knee injuries and padding on the back of the front seat. But these modest improvements soon fell by the wayside after Chrysler's "safety" engineers stated there was no way to test these modifications and, therefore, it was not possible to

82 THE SATURDAY EVENING POST March 13, 1948

Tucker Sets a New

128-inch wheel base.
Yet only 5 feet high from road to roof.
150-horsepower rear engine.
In the medium price field.

HOW TUCKER HELPS PREVENT TRAFFIC ACCIDENTS

TUCKER CYCLOPS EYE

▼

Sixty percent of traffic fatalities occur during dusk and darkness. Turning corners is especially hazardous because fixed headlights leave a blind spot around the corner. You can be in an accident before you can see it.

Tucker Cyclops Eye turns with front wheels, lighting your way around corners, giving you precious seconds to avoid accidents.

TUCKER SAFETY WINDSHIELD

▼

When impact hurls victim against conventional windshield, lacerations from broken glass or other injuries frequently result.

Tucker Windshield is safety glass mounted in rubber so that a sharp blow from within will eject it in one piece. Lacerations and skull fractures from hitting windshield are virtually eliminated.

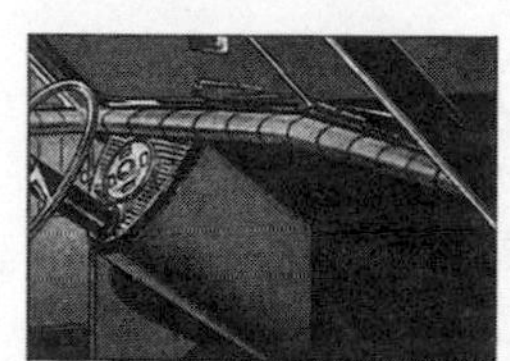

TUCKER CRASH COWL

▼

No. 1 hazard to front-seat passengers is the instrument panel, for a collision frequently plunges face or head against it.

The Tucker Crash Cowl of upholstered sponge rubber replaces the instrument panel which is moved to steering column below cowl. The Tucker crash cowl, like those in army tanks, reduces chance of serious injury.

TUCKER PRECISION BALANCE

▼

Rear-wheel skids when braking are frequently caused by sudden load shifts and faulty weight distribution. Weight of engine in front plus arrested impetus of car clamps front wheels down on the road, leaving rear wheels with less traction.

Tucker Precision Balance tends to eliminate this cause of skids. Rear engine adds weight on rear. So when brakes are applied car weight is more evenly distributed and all four wheels have more equal traction.

THE SATURDAY EVENING POST 83

Pattern of Safety

The National Safety Council says:

"Motor accidents kill or injure every 25 seconds of every day."

MORE AMERICANS were killed or injured in motor accidents last year than in all the battles of World War II.

Figure it yourself: Every 26 seconds someone is hurt by a car. Every five minutes there's an injury that permanently cripples. Every hour, four new customers for the morgue. *And remember* ... the next could be you or your children.

Is improved automotive engineering the answer to this tragic problem?

Of course, no car can be made so safe that reckless, careless drivers won't pay a penalty. But the new Tucker '48 can make the highways far safer for average motorists.

The Tucker '48 is the triumph of years of safety engineering... years Preston Tucker worked with the late Harry Miller building Miller Special cars for the Speedway, and with the armed services creating motorized vehicles for war.

This priceless experience will give you a car with a unique system of safeguards:

(1) Safeguards that make it far easier to avoid traffic accidents.

(2) Safeguards that give added security in unavoidable accidents.

(3) Safeguards that make driving so easy and free from strain that you can act on a split-second when an emergency occurs.

In a matter of months the Tucker '48 will be yours to drive, yours to marvel at the dozens of new features, every one exciting enough to be a major model change in a normal year.

You'll see a car longer and more luxurious than others in the medium-price field. Only five feet high from road to roof. Not 100, not 120, but 150 horsepower of smooth rear-engine power.

And what a joy to drive! No engine heat, fumes and noise flowing back through the passenger compartment because the engine is in the rear. Ordinary traction jolts and jars either eliminated or unbelievably softened by the exclusive new rubber torsional wheel suspension. A car, at last, with solid four-wheel stability, geared to the road.

In every way you'll see a completely new kind of performance in 1948's only completely new car!

Someone on your street may be among the first to own a Tucker '48. It could be you. A nationwide Tucker dealer organization has already been set up... will be ready to service your Tucker '48 wherever you are. Watch for dealers' announcements in local newspapers.

WHICH ALL MOTORCAR MAKERS ARE COMBATTING

TUCKER SAFETY CRASH CHAMBER

▼ The danger spot in a collision is the front seat, for in many cases rear-seat occupants can drop to the floor for protection. But front-seat occupants have no shield, and the impact may drive the engine back in their laps.

Tucker Crash Chamber, under cowl, is lined with sponge rubber and is protected by two steel safety bulkheads. Front-seat occupants can drop into it instantly.

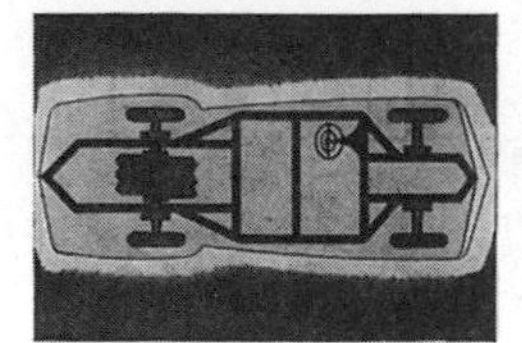

TUCKER SAFETY FRAME

▼ Many collisions aren't head-on but angle or side blows at points where often nothing protects occupants but the sheet steel sides and light superstructure.

Tucker Safety Frame surrounds passenger compartment, protecting passengers and vital parts of car. This frame is actually lower than center line of wheels, greatly reducing chances of overturning. It is prow-shaped to deflect angle blows with minimum damage.

Progress Report from the Tucker Plant

Right now the world's largest, most modern automotive plant is getting set to produce the Tucker '48. Forms and dies, jigs and fixtures are flowing in from suppliers.

Already the first fleet of pilot cars is being produced. Production lines are being set up. In a matter of months you'll see Tucker '48s on the road.

Tucker '48

Address Inquiries to
TUCKER CORPORATION, 7401 South Cicero Ave., Chicago 29, Illinois
Send export inquiries to Tucker Export Corp., 39 Pearl St., New York 4, N. Y.

retain them when considering the annual restyling of the car. Straith pressed on, and in 1946, found a sympathetic ear with Preston Tucker. Unfortunately, Tucker's new interior design quickly became a moot point when a series of political and legal machinations (some say orchestrated by the Big Three but never proven) forced Tucker into bankruptcy as the first of his fifty cars rolled off the assembly line—now lacking many of the advertised innovations in order to cut costs.

Tucker was not alone in his attempt to take on Detroit. The Kaiser-Fraser Corporation also had a go at the Big Three, producing cars from 1947 to 1955 with several interior safety innovations not offered in other automobiles. But these marginal attempts to challenge Detroit were soon overshadowed by a new option more enticing than any padded dash. It was horsepower. Before World War II, more horsepower was generally accomplished by increasing the size and number of pistons in the engine. Some prewar Cadillacs had as many as sixteen cylinders in a "V" configuration for smoothness and higher horsepower. More novel solutions included supercharging or compressing the fuel/air mixture before it was drawn into the pistons, providing a denser, more powerful charge of fuel and air to increase horsepower. But compression ratios (the amount the piston compresses the mixture before the spark plug ignites it) remained only 6 or 7 to 1 because of the low octane in gasoline. If this ratio were higher, the fuel would explode prematurely, putting added strain on the engine.

Charles Kettering was already retired three years from General Motors when he undertook his own experiments to refine a high-octane gasoline. The idea had been rejected by all of the oil-refining companies as impractical, but Kettering pressed on alone to find a way to increase the amount of compression the gasoline vapor could withstand without predetonating. Finally, in 1947, at the age of seventy-one, he achieved success and published a technical paper on his results, proposing a V-8 engine with a 12 to 1 compression ratio and overhead valves. The fruits of his labor were realized two years later in the 1949 Cadillac. Even though a large car, the new overhead valve V-8 got twenty miles to the gallon, but more importantly, the high-octane gasoline permitted a smaller, more powerful engine because of the higher compression ratio. The door was now open for a fuel-efficient motor that still provided excellent horsepower. But much to Kettering's dismay, his new breakthrough quickly caused engines to grow in size *and* power. With gas costing twenty cents per gallon and power-hungry options like air conditioning and power steering—not to mention the get-up-and-go needed for the highway—the horsepower race between the Big Three was born. Engines soon topped 300 cubic inches, heading for 350 and beyond. This was the age of the Dream Car: the mystique of power, comfort and speed behind the wheel. A willing marketplace was taking advantage of this seductive new image, especially in states that had no speed limits or states whose limits were not stringently enforced on their rural two-lane highways. The road was wide open and fast, and America was showing off its growing prosperity in ever longer and wider, chrome-laden ostentation.

Rural Two-Lane Highway
SPEED LIMITS
1954

STATE	SPEED LIMIT
ALABAMA	60 DAY, 50 NIGHT
ARIZONA	NONE DAY, 50 NIGHT
ARKANSAS	60 MPH
CALIFORNIA	55 MPH
COLORADO	60 MPH
CONNECTICUT	45 MPH
DELAWARE	55 MPH
FLORIDA	60 DAY, 50 NIGHT
GEORGIA	60 DAY, 50 NIGHT
IDAHO	60 DAY, 55 NIGHT
ILLINOIS	NONE
INDIANA	65 DAY, 55 NIGHT
IOWA	NONE
KANSAS	NONE
KENTUCKY	60 DAY, 50 NIGHT
LOUISIANA	60 MPH
MAINE	45 MPH
MARYLAND	55 MPH
MASSACHUSETTS	40 MPH
MICHIGAN	NONE
MINNESOTA	60 DAY, 50 NIGHT
MISSISSIPPI	60 MPH
MISSOURI	NONE
MONTANA	NONE DAY, 55 NIGHT
NEBRASKA	60 DAY, 50 NIGHT
NEVADA	NONE
NEW HAMPSHIRE	50 MPH
NEW JERSEY	50 MPH
NEW MEXICO	60 DAY, 55 NIGHT
NEW YORK	50 MPH
NORTH CAROLINA	55 MPH
NORTH DAKOTA	50 MPH
OHIO	50 MPH
OKLAHOMA	65 DAY, 55 NIGHT
OREGON	55 MPH
PENNSYLVANIA	50 MPH
RHODE ISLAND	50 DAY, 45 NIGHT
SOUTH CAROLINA	55 MPH
SOUTH DAKOTA	60 DAY, 50 NIGHT
TENNESSEE	NONE
TEXAS	60 DAY, 55 NIGHT
UTAH	60 DAY, 50 NIGHT
VERMONT	50 MPH
VIRGINIA	55 MPH
WASHINGTON	50 MPH
WEST VIRGINIA	55 MPH
WISCONSIN	65 DAY, 55 NIGHT
WYOMING	60 MPH

It is important to remember that during the early 1950s, the prevailing attitude of government, law enforcement and the general public was that it was the driver's

1951 Buick LaSabre with the director of the GM styling studio, Harley Earl, behind the wheel. © General Motors Corp. Used with permission of GM Media Archives

1958 Firebird III. © General Motors Corp. Used with permission of GM Media Archives

"Dream Cars" from the 1950s. While mostly a styling exercise of futuristic glamour and glitz, some of these cars dabbled with such innovations as anti-skid brakes and cruise control. The driving public, however, would have to wait decades to get such options on the standard American car.

responsibility to prevent accidents. There were virtually no governmental laws to mandate any safety performance standards for cars. What safety organizations existed (for example, the American Automobile Association and National Safety Council) were funded and influenced by the automotive manufacturers. When President Eisenhower held the "White House Conference on Highway Safety" in 1954, his executive order established the President's Action Committee for Traffic Safety, with Harlow H. Curtice as the first chairman. The fact that Mr. Curtice was also the President of General

Motors was not taken as the fox guarding the chicken coop. It was the logical extension of the influence, power and prestige held by the automobile manufacturers and the naive belief that they were serious about solving safety problems, of course, in human, not automotive, terms. Improved driver education and better roads came with no added cost to the manufacturers of automobiles. This mindset helped to influence the many changes Detroit went through in postwar America.

The pioneers who invented, built up and refined the automobile industry had recently passed on, with a new generation of men taking over. Walter P. Chrysler died in 1940 and Henry Ford and William C. Durant (the founder of General Motors Company) died in 1947. At about the same time, the automobile chassis reached, in mechanical terms, a state of "perfection." No longer would a company have to worry about designing new technology to ease operation or prevent breakdowns on the road. The automobile of the day was durable, dependable and relatively maintenance free. This left the facade of the automobile—its "styling"—to become the cost-effective way to entice buyers and stimulate sales. The fact that the American automotive manufacturers controlled 95 percent of the domestic market all but precluded the possibility of foreign competition bringing improved design innovations into the picture.

Along with this reality, a new breed of executive began directing Detroit's future automotive path. Schooled more in the areas of finance than engineering, these men understood the need to control automotive costs better than the necessity to explore innovations. When Henry Ford II and his team of "Whiz Kids" took over Ford Motor Company during and just after World War II, their financial acumen rescued the company from bankruptcy and returned it to profitability. This movement towards bean counters running the Big Three had phenomenal success at cutting costs and maximizing profits. But this form of management sacrificed innovation and reinvestment in the company itself. With financial gain clearly placed over product development, the automobile itself became an abstraction at the upper levels of management. It became one more "thing" to sell. This executive detachment from the product prevented the auto industry from making basic design changes to improve safety.

Image and horsepower sold cars instead. A larger engine and fresh exterior styling didn't require the complete overhaul of the frame and running gear. It cost less money over the short term, too.

With Detroit's new attitude towards the automobile, and the general feeling that the driver was responsible for preventing accidents, the car companies enjoyed a certain immunity from criticism about their products—except in the area of horsepower. The growing horsepower race was considered by many in the highway safety community to be a dangerous trend that should somehow be stopped. Even Edgar J. Kaiser, president of the Kaiser-Fraser automobile corporation, denounced the race as "ridiculous." But given the power and influence of the Big Three, along with the desire of the driving public for faster acceleration, such criticism had little effect. Grumblings from the National Safety Council, the insurance industry and law enforcement agencies lacked the organizational and national resolve to slow down the horsepower race, let alone bring it to a halt. The only possibility seemed to be to raise a public stink about the problem in the media. This caused the car companies to back off of their more overt advertising for brief periods, only to have the ads reappear again at some later date. The industry was also careful to *suggest* speed and power rather than show it outright. A foot tromping the gas pedal or a car streaking down the road at some undetermined speed sent an implied, but clear, message to prospective buyers about horsepower. "Rocket" V-8s, jet-planelike tail fins and a reassuring male voice extolling the virtues of "newer, bigger, improved engines" enticed customers into the showroom. Prospective buyers squealing the tires and being pushed back into the seat on a test drive did more to sell cars than high-pressure salesmen. Everyone knew what was going on, but little, if anything, could be done to stop it.

Since this assault on the problem was yielding no results, the next action was to decry the by-product of horsepower—the evils of speed—and incite the outrage of the public and law enforcement to do something about it. There was growing concern for all the death and suffering on the road which had now reached 35,000 fatalities a year. "Slow Down the Speeder" became the mindset of the 1950s. This was a sporadic campaign at best. Using the local gendarmes or state patrol, a temporary enforcement blitz would hit county or state roads with a vengeance, trying to slow down speeders to something near the posted limit. This generally succeeded in reducing speeds in the immediate area for as long as the police were around, or at most a few hours afterwards. This action would send hundreds of motorists home licking their wounds, with a fine and a hard-luck story to tell their friends. Troopers would earn their keep for force and state, using techniques that ranged from pacing themselves behind the speeder in question, to laying down rubber hoses which recorded the speed as vehicles drove between them. Airplane patrols became more common, timing cars between marks painted on the highway, radioing to a squad car down below. Radar, in use since 1948, was becoming more popular, but it required a substantial setup, and worked only in a stationary mode that demanded delicate leveling of the radar unit and monitoring of a printed graph. A second chase car was also needed to apprehend the guilty party. These enforcement binges garnered plenty of media attention but did little to reduce the overall speed of traffic or lower the total number of people killed.

The most famous speed enforcement campaign of the time took place in Connecticut in 1956. Governor Abraham Ribicoff declared war on speeding, creating an unparalleled enforcement blitz on the state's highways. Thousands of signs warned against speeding, and the threat was backed up with a yearlong campaign to enforce the posted limit. Driver's license suspensions rose from 374 in 1955 to 10,055 in 1956. During this time, the number of accidents and injuries *increased* and the state's death rate for traffic fatalities remained higher than the national average. The campaign succeeded in raising the ire of most drivers on the road, leaving a legacy of futile speed enforcement that Connecticut and many other states did not heed. Advocates of these enforcement blitzes maintained that if they could just get enough police out on the road, motorists could be "encouraged" to obey the limit. Privately, many state and law enforcement officials scoffed at these slowdown campaigns but marveled at the amount of revenue speeding tickets could generate—all with the "cooperation" of the driving public.

New York State, 1957. The "warning" virtually no one heeded, making it useless. Monty McCord - Police Cars: A Photographic History

Beyond the fervor and acrimony about speed, horsepower and safety—a seemingly endless debate to be recirculated in the media with no resolution—America's infatuation with the car and driving continued to grow unabated. One problem needed swift attention, however, or the love affair would be in serious trouble. The glut of traffic was running out of sufficient roads. The "open" road had become anything but in many parts of America by the mid-1950s, and mounting pressure to do something about it had reached critical mass. Since the late 1930s, there had been much talk, and little else, about a network of national expressways to connect the country and speed commerce. Visions of Futurama-like freeways and the growing number of private or state-run turnpikes, mostly in the East, generated much excitement but little money to ease the problem. Consensus for a specific plan seemed nonexistent. A whole host of special-interest groups fought over the type of road and taxes (or lack of taxes) to benefit their particular needs. Farmers' groups, the trucking industry, petroleum distributors and auto owners associations each fought the other for the best farm-to-market road funding or for reduced federal gas and vehicle taxes to aid their businesses.

By 1944, Congress did manage to authorize the construction of a national system of expressways 40,000 miles long but neglected to appropriate the money to build it. The battle dragged on for another twelve years, tolerating all manner of infighting and political intrigue while the driving public became more enraged by an intolerable road problem. Finally, in June 1956, President Eisenhower signed The Federal Aid Highway Act which, in turn, created the Highway Trust Fund. This monumental piece of legislation would provide 90 percent federal funding for a National System of Interstate and Defense Highways 42,000 miles long. The "Defense" title was tacked on to build more support, taking advantage of Cold War paranoia and the need for military mobilization to protect against a Communist sneak attack. Paid for by the users of the highway in the form of a four-cent per gallon tax on gasoline and excise taxes on tires and lubricating oil, this $27.8 billion, twelve-year freeway construction program would become the greatest public works project in the history of the world.

With a traffic jam of 65 million cars, trucks and buses already on the road (up from 31 million in 1945), the interstate system seemed like a dream come true. *Time* magazine declared, "Motorists will be able to drive from Los Angeles to New York over the federal network without passing a single traffic light or intersection." An article in *Holiday Inn* magazine stated, "It will be possible to speed along at 70 to 80 miles an hour on these superhighways." With average highway speeds for passenger cars hovering around 50 mph, this only bolstered expectations for the freeway network to come.

The coming of the interstate system brought with it a slight, but only temporary, shift in attitude towards high-speed driving within the safety community. The right-of-way for the new freeway created an even more spacious roadway than the one laid down by the Pennsylvania Turnpike sixteen years earlier. The lanes were still twelve feet wide but the shoulders, and more importantly, the center median strip were now two to four times wider on most rural stretches, allowing for a greater separation of cross traffic and less likelihood of a head-on collision. The space underneath overpasses was at least fourteen feet high and wider from the edge of bridge abutments to the shoulder. This seemingly endless ribbon of highway with its leisurely curves and easy rises meant all-day, worry-free driving at high speed. Estimates at the time expected the new interstate to save 3,500 lives annually, reduce the cost of accidents by $725 million and save another $825 million for truckers by cutting travel time, fuel use and wear on tires and brakes.

The old system of two-lane highways winding its way around hills and through towns was a fatiguing, often backbreaking way to travel great distances. It was also filled with plenty of peril. My father would quite often travel on business by car from Minneapolis to Chicago. Trying to make time by driving faster than one should between small towns (Wisconsin's rural speed limit then was 65 mph during the day, 55 mph at night), he still needed eight to ten hours to cover this 410-mile trip, given speed traps and unexpected detours. Once, at the crest of a hill on an empty stretch of two-lane highway, two semis appeared lined up grille to grille, and hell-bent on overtaking each other. With only seconds to decide, my dad took the ditch at over 70 mph, threading his way through a row of telephone poles, using the car's momentum to skid sideways up the ditch and back onto the road. In 1950s America, this was an all-too-common experience for many drivers—with not nearly so lucky results. With the opening of the interstate, this type of occurrence was all but impossible. My father's travel time was cut to six hours on a good day.

This revolution in road design, which created a smooth, uniform strip of freeway with limited-access points and no at-grade intersections, helped to reinforce the notion that the road, not the car, would promote safer driving. But within certain circles, there was a growing body of evidence that most accidents, even at high speed, were survivable if the driver and passengers were protected inside a properly designed vehicle. A thorough, scientific investigation of this began during America's involvement in the Korean War when the armed services discovered, much to their dismay, that they were losing more soldiers on the highway than on the battlefields of Korea.

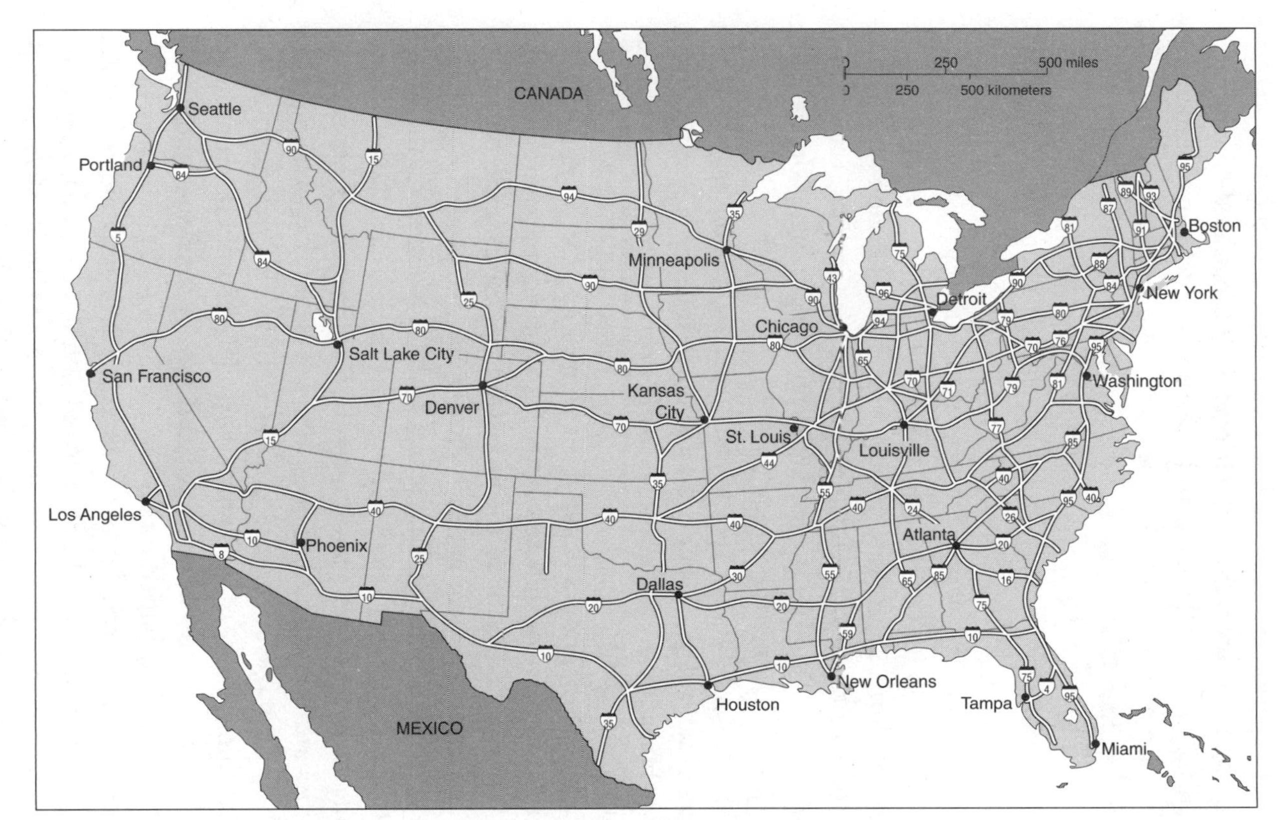

The greatest public works project in the history of the world. The National System of Interstate and Defense Highways—45,000 miles long.

Calling upon Air Force Colonel John P. Stapp to research the physiological components of a crash, the military set out to undertake a series of experiments. A Ph.D. in biophysics and an M.D., Stapp had several years of research on the forces the human body could tolerate during aviation crashes and ejection from fighter planes. His findings were also applicable to the sudden deceleration in automobile collisions. He concluded that if properly restrained with a harness and protected from striking any objects, the human body could endure the shock of deceleration from speeds as high as 600 mph. If unrestrained, a person could be killed striking a sharp object as slowly as 5 mph. Colonel Stapp was no stranger to such experimentation, volunteering to ride a rocketsled up to 632 mph and to be slammed to a halt in just 1.4 *seconds*. Though Stapp's quick stop was some eighteen times longer than the average car crash, it was calculated that the Colonel endured the same force of deceleration as if he had driven a car into a brick wall at 60 mph.

Despite national exposure in the press, and the development and testing of several passenger-friendly vehicles, Colonel Stapp's experiments brought no changes in Detroit's car design. However, such research did not prevent the Big Three from discreetly hiring an individual or creating a department to monitor the development of automotive safety outside the industry. Although the outcry for safer cars was still small and coming mostly from the medical community, it was slowly spreading to such diverse places as the insurance industry and the halls of Congress. By 1956, hearings on the broader subject of highway safety had become more focused on the automobile itself, now that a substantial body of evidence from research facilities attached to universities and the military was being made available, though not to the general public. This information was kept within the research community in deference to the automobile industry, whose power, influence and money kept this knowledge, for the most part, buried.

It would be incorrect to assume that the car companies were totally devoid of conscience. There were a handful of visionaries who felt they had a responsibility to improve automotive safety. Alex L. Hayes, safety engineer, and Robert S. McNamara, the assistant general manager of Ford Motor Company, were two such men who had this foresight. Pragmatically, McNamara reasoned a modest safety package offered in Ford cars might offset any notions of future government intervention in mandating safety. Such action could set a trend that the other manufacturers would follow. Furthermore, Ford's sales against the snappier Chevrolet with its more powerful, small-block V-8 looked even more bleak in 1956 than in '55, since Ford had applied only minor cosmetic changes to its lineup. McNamara was convinced that selling safety was a whole new, untapped area for increased sales. Lobbying hard to get the go-ahead from a skeptical corporate management and board of directors, he was finally able to convince them that the growing fervor within and outside government could be quelled with this safety campaign. Time constraints and assembly line lead-time held back the program to a more add-on endeavor of dash and sun visor padding, seatbelts, "safety" door latches and an energy-absorbing steering wheel. The advertising was more elaborate than the safety package itself, appearing in newspapers, magazines and on a four-minute commercial during Ford Television Theater. A national survey showed that 60 percent of all car owners grasped the basic premise of what the safety package was supposed to do and that it was associated with the Ford Motor Company. At the Chicago Auto Show, 31 percent who expressed interest in buying a new Ford gave safety as the main reason. The safety community was justifiably impressed, hailing Ford's entry into passenger protection as a step in the right direction.

Glowering down with great displeasure at this burgeoning interest in auto safety was the executive management at General Motors. Using safety to improve sagging sales was bad enough, but Robert McNamara seemed to be genuinely out to improve automotive safety at Ford. With the prospect of McNamara someday running the company and further influencing safety innovation (he later did become president), this was something that had to be nipped in the bud—*now*. Garnering 55 percent of the market share at the time, General Motors was quick to point out to its competition—with only slight exaggeration—that Chevrolet could lower its price $25 per car to bankrupt Chrysler and $50 to bankrupt Ford. When the edict came down from GM, Ford backed off its safety campaign. Several upper-level Ford executives were former GM employees, so

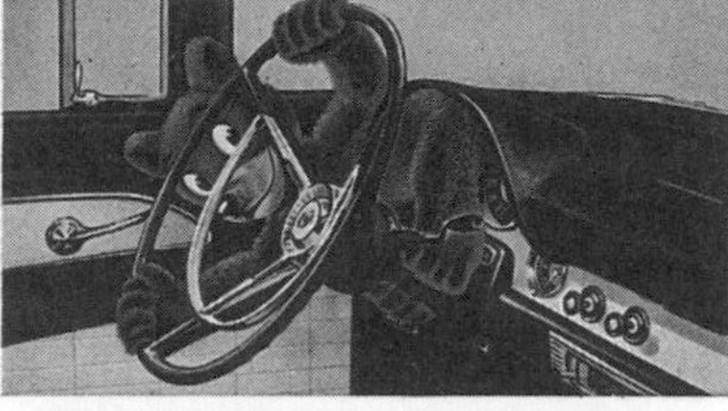

This Lifeguard steering wheel has a new deep-center construction to reduce the possibility of driver being thrown hard against the steering post in case of an accident. Our Ford engineers have mounted the rim of the wheel high above the recessed steering post to help "cushion" your chest against severe injuries from impact.

Optional Lifeguard padding protects you against accident injuries by providing a "crash cushion" on both the instrument panel and sun visors. It is five times more shock absorbent than foam rubber. New Lifeguard double-swivel rearview mirror that "gives" on impact and resists shattering is standard on all '56 Ford models.

You're twice as safe if you stay inside the car in an accident. Statistics prove it conclusively. So our Ford engineers have designed these new Lifeguard door latches with a *double grip* to reduce the possibility of doors springing open in a collision.

Look at this Ford seat belt! One-third stronger than required for airlines, it is securely anchored to reinforced, all-steel floor structure. Optional Ford seat belts can be adjusted or released with one hand . . . are available in colors to harmonize with interiors.

FORD V-8

Sells More because it's Worth More!

Courtesy of Ford Motor Company

it came as no surprise when Ford's chairman of the board, Ernest R. Breech (a former chief financial officer at GM), told the advertising department to shift its emphasis from safety to styling and performance. This came at considerable expense to Ford, which had to switch its planned-for advertising campaign away from improved auto safety.

The end result was that Chevy outsold Ford in 1956, fostering the misguided belief that safety would not sell. The forgotten reality of this campaign was that 43 percent of '56 Fords were ordered with optional safety padding—the highest percentage for any option, even beating out the automatic transmission—and one out of seven customers ordered seatbelts. Furthermore, as the years passed, these minimal improvements to the interiors of Ford cars helped to reduce injuries to the chest by 50 percent for drivers who struck the deep-dish steering wheel, and the safety door latches prevented doors from popping open by an improved 60 percent. Even though the safety package was not dropped, it was never again advertised. The driving public showed that it was interested in some kind of safety, but for the time being, General Motors was determining just how strong that interest would be.

The question of how to design a more crashworthy automobile, or more importantly, how to build one if starting from scratch, did not come from inside the automobile industry, as one might expect. It came instead from an insurance company. Liberty Mutual, working in consort with the Cornell Aeronautical Laboratory, designed and built a nonoperational safety car in 1957. First proposed in 1952 by Frank J. Crandell, Liberty Mutual's chief engineer, Survival Car I had many innovations that would allow the driver and passengers to walk away from a collision of up to 50 mph. From the outside, the car appeared more or less normal by the standards of the day. The interior, however, took on a radical appearance that was different from anything else on the road. The doors folded open telephone booth style with a seating arrangement that was unconventional to say the least. The driver sat front and center with passengers on either side. A backwards-facing passenger sat directly

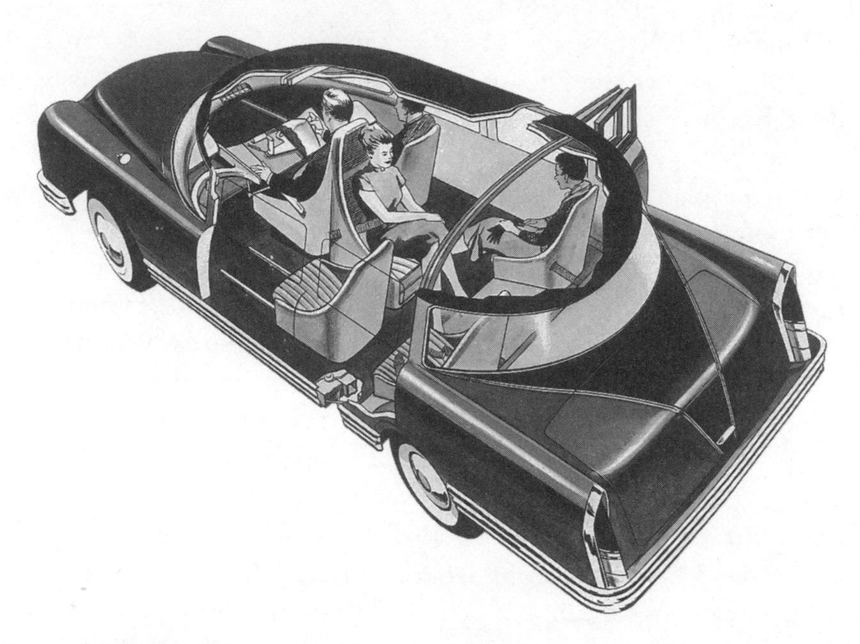

State-of-the-Art Safety, 1957 style. Liberty Mutual-Cornell Survival Car I. The car that would never be. Liberty Mutual

behind the driver with two more facing forward to the left and right. This provided all occupants with their own "safety space" in the event of a crash. The driver had a 180-degree, unobstructed view of the road with a windshield that curved around the front of the car. Hydraulic levers, not a steering wheel, controlled the vehicle. Based on extensive research by the Cornell team of engineers, Survival Car I clearly showed that great strides could be made in the field of better passenger protection. And the public and media did receive it with a favorable amount of fanfare. Detroit, on the other hand, considered the radical design changes for the interior too impractical or, at least, too far ahead of their time to be acceptable.

This rejection of the first prototype by the industry did not deter Crandell and the Liberty-Cornell team. To show that safety was not some far-off dream of the future, they modified four 1960 Chevrolets, installing a collapsible steering column, separate front and rear brake systems and a reinforced passenger compartment with rollover bars. The most visible change to the car was the removal of the standard bench seat, which was replaced with "capsule chairs"

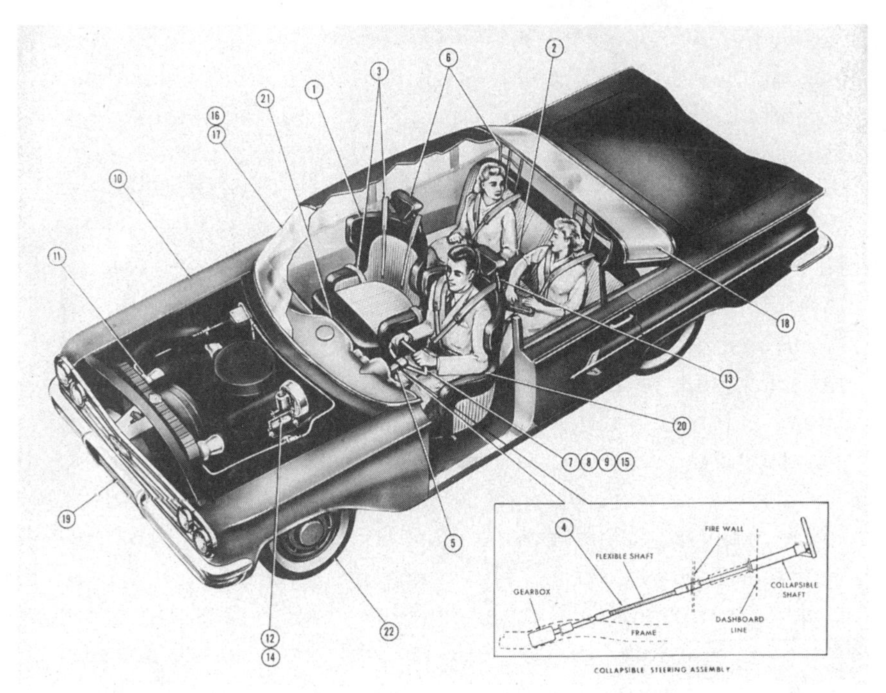

1) Capsule chairs, so designed and so tied to floor as to restrain driver and passenger in a 30-mph collision against a 5000-lb blow and a deceleration force of 30g in a front, rear or lateral crash

2) Protection of passengers in rear seat against a 5000-lb blow

3) Restraining lap belts and shoulder harnesses on all seats

4) Flexible steering shaft that will buckle in a crash situation

5) Steering tube that telescopes a total of 8 in. during a crash

6) Whiplash protection on all seats to restrain head in event of rear-end crash

7) Rectangular steering wheel to prevent breaking or bruising kneecaps

8) Steering wheel of reduced diameter for greater visibility

9) Improved maneuverability, faster turning ability and greater driving stability through utilization of special driving wheel

10) Unit body construction with high energy-absorption factor that will collapse and ruin car at moment of severe impact but protect passengers from bodily injury

11) Automatic fire-control system utilizing carbon dioxide

12) Safety brake device

13) Rollover bars in capsule chairs for greater head protection

14) Power brakes

15) Power steering

16) Laminated safety windshield with double-weight filler to afford greater resistance to penetration with no heavier blow upon head

17) Saflex interlayer windshield to eliminate 95 percent of ultraviolet rays

18) Tinting of all glass windows reduces heat load normally entering car through transparent glazing by approximately 30 percent

19) Reflective license plates for greater visibility at night

20) Support arms to reduce driver and passenger fatigue

21) Alert-O-Matic signal system to awaken driver if he falls asleep or stops car if he does not awaken

22) Micro-siping of tires to increase traction and reduce skidding possibilities on wet, icy and snow-covered roads

Additional features are side reflection mirrors for greater driver visibility, and smooth hood over engine to reduce injury hazards to pedestrians, if hit.

Liberty Mutual's Survival Car II. Affordable safety. Liberty Mutual

that wrapped around the driver and front passenger with a web of steel and padding. Complete with 4-point seatbelts, headrests to prevent whiplash and a swivel base to make getting in and out of the car easier, these chairs were designed to protect a person against a front-, side-, rear-end or rollover collision at 30 mph. These, and a host of other small improvements, demonstrated that Liberty Mutual's Survival Car II could make available to the driving public affordable, comprehensive passenger protection for crashes at modest speeds. Again, the automotive industry, for the most part, remained mute to these safety improvements. But the president of General Motors, John F. Gordon, did express his animosity while speaking to the National Safety Congress in 1961: "It is completely unrealistic even to talk about a foolproof and crashproof car. This is true because an automobile must still be something that people will want to buy and use." This cynicism about safety and Detroit's denial of a substantial body of evidence showing passenger protection being not only possible, but desirable was, on the one hand, a display of the absolute arrogance and contempt—especially in the case of General Motors—held by the American auto industry for its customers' safety or for anyone who criticized the cars it built. On the other hand, the industry did grasp what the people would or would not buy. Moving safety forward in such a revolutionary step as Survival Cars I and II only served to magnify the gap between what was possible and what little the industry was doing to promote better safety design. At the time, front seats like those installed in Survival Car II would have been an extremely tough sell, despite their obvious advantages. Bucket seats were unknown to the average American and the industry immediately focused on the different appearance of what was, then, state-of-the-art safety—never bothering to look at its own products to see how unsafe they really were. That would take a few more years.

The growing safety movement in this country was becoming more preoccupied with the automobile's lack of "crashability" and its nonexistent passenger protection in a collision. Many members in the various safety and law enforcement organizations failed to realize that the standard car of the day was a nightmare of poor handling and braking. They continually griped about the ever growing amount of horsepower under the hood, but did not recognize that this larger engine was being placed on the same low-grade tires, soggy suspension and anemic brakes that were now required to do more work to keep the car on the road. Driving down a new stretch of interstate with its subtle curves and rises posed no problem, even at speeds of 70 to 80 mph. But try to come to a quick stop, or turn to avoid some obstacle, especially during the first critical seconds before a potential accident, and now you have a real crash with serious consequences. In the early 1960s, the magazine *Popular Science* sponsored a comprehensive brake test for various makes and models. There was a tremendous discrepancy in braking distance between cars, ranging from 170 to 359 feet in a panic stop from 60 mph. (The average car today can stop from 60 mph in the range of 125 to 160 feet.) At the speed many cars were being driven in those days, this was a serious safety concern. Not only was there little passenger protection, but there were hidden dangers when a car was forced to turn or stop, even at modest speeds. To experience this combination of power and panic is, to say the least, nerve-racking; the difference between then and now is an important distinction for those of us familiar with modern tires, brakes and suspension.

I once had the debatable good fortune to climb behind the wheel of a friend's 1963 split-window Corvette. This was the one with the more moderate 327 cubic-inch, 350 horsepower engine mated to a 4-speed Hurst shifter. Off to have a go at an abandoned country road, the Vette surged down the street with a power and torque that pinned you back in your seat, putting a big smile on your face. A quick flick of the shift lever and a pop of the clutch would easily chirp the tires going up the gears. This is what it was like, driving in America in the early 1960s. The exhilaration of raw, brute horsepower and not a care in the world—until you had to stop. Climbing the gears and numbers on the speedometer, the Corvette and I shot down the wide-open country road at just over 100 mph. Approaching the only turnaround within the next mile, I hit the brakes and sailed right on by. The four-wheel drum brakes were so bad, I had to wrench on the steering wheel, strain my back into the seat and use both feet on the brake pedal to haul the car down to a more modest speed. If it had been a true emergency, I'd have been a goner. I imagined, after it was all over, that the sharp tingle in my stomach and up my spine was probably the last feeling many drivers had on the road in the early '60s.

The reality of that tingle, however, was the furthest thing from the minds of most young American males as they lined up to buy the latest "muscle cars" coming out of Detroit by the mid-1960s. Such automobiles were first modified by young hot rodders who wedged huge motors into smaller cars, installing tighter suspensions, better brakes and wider tires to cope with a massive amount of horsepower crammed into a lighter and, now, much faster car. Drag racing on tracks *and* city streets became commonplace on weekend evenings as youthful drivers burned rubber from one street light to the next, cruising down their favorite strip of drive-ins, movie theaters and gas stations. In my hometown of Minneapolis, it was Lake Street that drew these illegal young drag racers and carloads of teenagers to spot for cops, so challengers could screech it out with the rumored champion. Every large town had its "strip," the most famous being Woodward Avenue in Detroit. Here was the official birthplace of the muscle-car phenomenon (as far as the Big Three were concerned). This impromptu drag strip spurred several people within the industry to capitalize on this trend, supplying the growing number of baby boomers hitting the road with a series of high-horsepower supercars that would scorch the pavement right off the showroom floor.

One such chief engineer to exploit this market was Pontiac's John Z. DeLorean. Assembling a team of advertising and engineering people, he took the robust, 348 horsepower "Tri-Power" V-8 out of the full-sized Bonneville and shoehorned it into the smaller and lighter-weight Tempest. Pontiac borrowed a certain three letters from a much beloved Ferrari model and the GTO was born. These factory muscle cars eliminated the need for young motorheads to modify their own automobiles further (even though they did), and cars like the GTO helped to create one of the strangest times in American automotive history. An entire culture sprang up around the phenomenon. Celebrated in Beach Boys' music, teenage slang and ritual rites of youthful passage, phrases like "4-0-9," "hemi" and, of course, Carroll Shelby's "Cobra" worked their way into everyday conversation and, finally, history. "Horsepower" was on the lips of every red-blooded American male. It became a form of worship, genuflecting to the power, speed and exhilaration these muscle cars could bring to mere mortals. It was the absolute pinnacle of our indulgence with the automobile.

As a small boy, I was given a ride in my cousin's "goat" as they were affectionately called in those days. He took great delight in spinning the GTO's tires at every stop sign, pulling away so quickly that I was pinned down in the seat next to him, unable to move my arms or body, the g-force was so great. Though both of us managed to live into middle age with many fond memories of those crazy days, some horsepower worshipers did not. Safety was something you thought about the split second before you met your Maker, or it was the afterthought of those gathered around your headstone. To admit *this* was to deny all the desire, pleasure and thrills horsepower brought to those celebrating its virtues. But the sad and unspoken reality remained: The grim reaper was also a willing participant.

There was another small but growing group of speed and power enthusiasts out on the road. It was the highway patrol. Arming themselves with high-performance squad cars, troopers in "Interceptor" Fords, "Police Pursuit" Dodges, and Plymouths with the "Golden Commando" V-8 prowled the highways and interstates looking for the more flagrant highballers to chase down and issue a speeding citation. A curious symbiosis developed over the years as car buyers craved more horsepower and speed, and the highway patrol followed right behind in hot pursuit. Engines soon topped 400 cubic inches and 400 horsepower. This "game" between the pursuer and the pursued almost always ended with the driver being pulled over by the trooper. In general, these highway patrol officers were more familiar with driving at the upper limits of their squad cars which, fortunately for them, were equipped with heavy-duty suspensions and tires with safety innerliners to maintain control while slowing down if they had a blowout. "Speeders" rarely, if ever, had these rudimentary tire and suspension improvements but, in ever growing numbers, were taking advantage of the speed and power offered by Detroit.

THE HUNTERS . . .

1963 Dodge 880

1962 Chevrolet (Could there be a 409 V-8 under the hood?)

1963 Ford Galaxie

For hot pursuits in the early 1960s, squad cars from Ford, Chrysler and General Motors produced anywhere from 325 to over 400 horsepower. When fighting the law, the law almost always won. Photos by Cpl. Ed Sanow

. . . AND THE HUNTED

1964 Dodge 426 "Hemi"

1965 Pontiac GTO

1963 427 Ford Galaxie

Rogues' Gallery circa 1964–65. These cars could pull a 1/4-mile time in 12 to 13 seconds at over 100 mph stock-out-of-the-box. Photos by Alex Gabbard

Most were oblivious to the potential danger no further away than the low-grade tires working hard to burn up the pavement.

As more miles of interstate were opened for travel, speed was increasing on those routes along with the grace given by troopers patrolling the road. By the mid-1960s, only Montana and Nevada were still under "reasonable and prudent," which basically meant no speed limit for all who drove the rural highways in those states. Montana did have a nighttime speed limit of 65 mph for the interstate and 55 mph for all other highways, enforced rigorously by the few troopers who were out on the road. For most states, the posted limit for the freeway was 70 mph, with a few at 75 mph and even 80 mph for the Kansas Turnpike. Minimum speed limits also began to appear on many stretches of interstate, and troopers began pulling over slowpokes in an attempt to eliminate the potential danger of slow and fast drivers meeting up in an untimely fashion. An older friend of my father's remembered being given a stern lecture by a trooper who, in no uncertain terms, explained to him the virtues of not dilly-dallying on a freeway that was built for speed. During this time, it was a "speed demon's" paradise, and many drivers were forsaking the more traditional highways like Route 66 for the free speed of the interstate. On a 70 mph freeway, the average state trooper of the day did not pull people over until they were over 80, or in some cases, 85 mph. This put the "speeder" well over that number, which was no problem for the ever larger engines coming out of Detroit, but increasingly risky for the poor tires, vague brakes and handling that became progressively worse as car weight, engine size and speed increased.

There was another unpleasant fact which became apparent as the annual death toll, now over 40,000, continued to increase. About 3,000 of that number were dying on the interstate system, which was about one-third complete. This put a certain amount of tarnish on the image of a superhighway that was supposed to be the ultimate in safe, high-speed travel. Safety experts and traffic engineers were quick to point out, however, that the tremendous increase in traffic would have caused a much higher death toll if the freeway system had not been built. This diverted the questions and obscured the real answers to the problems of why this new road was not a safer mode of transportation. Over the following years, an improvement in the freeway's roadside safety was undertaken to remove more obstacles and to shield potential dangers such as bridge abutments with guardrails and energy-absorbing barriers. In some cases these were misplaced, causing the very accidents they were designed to prevent. What to do about the end of the guardrail itself became a subject of debate. Most states preferred using the "Texas twist," turning the end of the guardrail down and anchoring it to the ground. This was a trade-off at best. It prevented the guardrail from crashing through the windshield when a car struck the end of it, but did not stop a car from launching itself into midair when the end was twisted into a ready-made guardrail ramp. "Shear-away" street lights and signs also helped to minimize death and suffering. But the question of speed's relationship to road safety would not go away, so the federal government set out to investigate the problem.

The first major study to be published was "Accidents on Main Rural Highways Related to Speed, Driver, and Vehicle" by David Solomon in 1964. Analyzing the accident/speed relationship from a substantial body of information collected in eleven states, Solomon's study reviewed the accident records of almost 10,000 drivers, and included interviews with 290,000 other motorists who used the roads studied in his report. Starting in 1957, looking at two- and four-lane, noninterstate highways, the study produced a wide variety of conclusions that confirmed some preconceived notions and dispelled others:

- Accident involvements ". . . were highest at very *low* speeds, lowest at about the average speed of all traffic, and increased at very high speeds, particularly at night."
- "The severity of accidents increased as speed increased, especially at speeds exceeding 60 mph."
- "The fatality rate was highest at very high speeds and lowest at about the average speed."
- "Passenger-car drivers under twenty-five years of age and more than sixty-five years of age had the highest involvement rates."
- "Local drivers tended to have higher involvement rates than other drivers, particularly at night."

• "Drivers of passenger cars having low horsepower had *higher* involvement rates than drivers of cars having higher horsepower, regardless of the other variables studied. This may be related to the relatively poor acceleration capability at highway speeds of cars having low horsepower."

The study predates the advent of muscle cars and these high-horsepower vehicles never did make up more than 10 percent of the car market. The bulk of this information, when plotted on a graph, creates a U-shaped curve that represents the accident/speed relationship. With the accident-involvement rate on the left side and variation from the mean speed on the bottom, this graph shows that motorists driving much faster or slower than the average speed of traffic have a higher accident-involvement rate, and that large variations in speed can increase accident risk. Drivers just above the mean speed, however, have the lowest involvement rate. If, for example, the average speed were 60 mph, those traveling much slower or faster would have greater risk of an accident, but those traveling 5 to 15 mph faster would be at about the same level of risk. The point where the accident curve rises noticeably upward coincides with the speed that 85 to 90 percent of the traffic flows at. These findings confirmed the practice of traffic engineers setting speed limits at the 85th percentile mark. Since the 1930s, this level had been used to treat the majority of motorists as law-abiding citizens. Now, it could be set as the limit for "safe" driving.

Subsequent federal studies undertaken by Julie Cirillo of the U.S. Department of Transportation revealed a similar, but deeper curve with lower involvements for interstate freeways, corroborating most of Solomon's study. None of these studies changed the fact that a growing number of people were dying on all roads, including the new interstate. It did help to substantiate that the freeway, by design, reduced accidents and that speed limits in the range of 70 to 75 mph were within the acceptable range of the U-shaped curve's "tolerance." Federal speed estimates for the interstate showed a 60 mph average at the time. Adding 10 to 15 mph from that mean-speed put the "official" speed limit in the correct area. With average speeds increasing 1/2 mph per year, regulation of speed on the interstate appeared to be moving in the right direction for continued "safe," high-speed operation. Unofficially, the grace given by the average state trooper and the growing thirst for power and speed by many drivers put more and more of them at increased risk, whether they realized it or not.

About the same time the Solomon study was published in 1964, two cars were introduced which were to have quite an effect on the driving public. The first one started out as a prototype that promised many new and exciting innovations. Displayed at various points around the country, this experimental car from Ford showed what the industry already knew about safety design and what it could do with that knowledge. The car sported bucket seats with lateral holding power for hard corners, fixed rigidly to the floor pan of the car and designed with headrests to prevent neck whiplash in the event of an accident. Lap belts and optional shoulder harnesses were provided along with a roll bar, collapsible steering column and "fail-

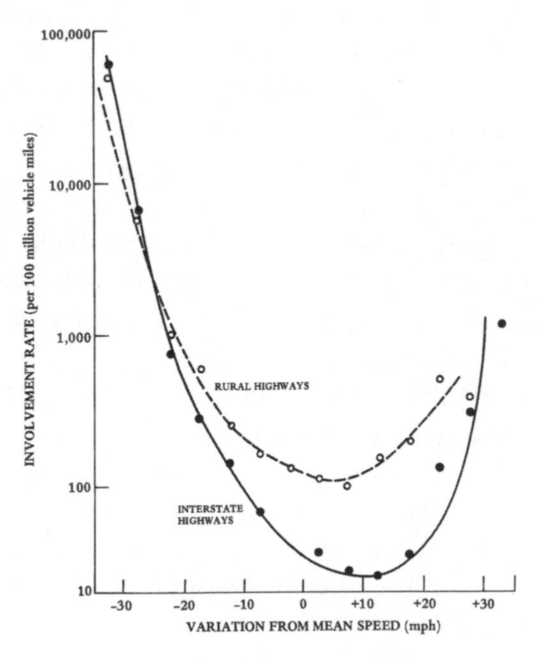

Accident-involvement rate by variation from the mean speed.

safe" dual-circuit brake system. When the production version of this prototype—the Mustang—debuted in April 1964, Ford decided not to retain any of these safety features on the new car. In the next twelve months, the Mustang would set outstanding sales records, putting nearly half a million on the road. Had these safety features been retained, this car could have set a dramatic safety precedent on the American highway.

The other car had already been on the road a couple of years with modest sales success. This sporty little import fighter was manufactured by the giant in the industry, General Motors. With four-wheel independent suspension and a rear-mounted air-cooled engine, this was the most technologically advanced car coming from Chevrolet in years. The Corvair, however, had one fatal flaw in its design. The rear suspension was so poorly engineered that, under certain circumstances (common enough on the road), the driver could lose control and have an accident simply from this inferior design. Having only one universal joint for each of the two driveshafts, when the tail-heavy Corvair took a hard corner or hit a severe bump, the swinging action of the wheel and driveshaft—moving as a single unit—would cause the tire to have less contact and adhesion with the road, inducing loss of control. This design flaw was the impetus that started General Motors down the road to decline. Now, many who had for years criticized the lack of auto safety—and who were bruised from previous battles with the industry for things as simple as safety belts in cars—banded together in and outside the federal government to take the Big Three to task.

If the Corvair had provided the impetus, then challenging a giant like the auto industry would require some kind of catalyst to help set everything into motion. That catalyst was a young lawyer named Ralph Nader. Though many other factors played into this particular assault on the makers of automobiles, Nader simply had to draw upon the extensive body of automotive safety research already available and tell the truth. Had this attack on the American auto industry taken place at another, less-prosperous time (prosperity, incidentally, which was directly caused by the auto industry itself), its impact would not have been as noticeable. Most of us were comfortably affluent in the mid-1960s, feeling confident about our jobs and pocketbooks. The government was moving into a more aggressive stance towards social involvement, and the auto industry was mired in a spoiled and arrogant attitude of supreme power, regarding any intrusion into its domain—no matter how good or righteous—as an opportunity to show flaws in the accuser.

At this unique moment in time, a group of men came together in Washington to summon the mighty titans of Detroit before them. Abraham Ribicoff, the former "speed kills" governor of Connecticut and now a U.S. senator, chaired the subcommittee on Executive Reorganization of the Government Operations Committee. He had become interested in the concept of automobile passenger protection after the failure of his speed enforcement campaign back in 1956. Ribicoff contacted Daniel Patrick Moynihan, the Assistant Secretary of Labor for Policy Planning since 1962 and author of the watershed article on safety, "Epidemic on the Highways." Working as a consultant on auto safety to Moynihan was Ralph Nader, who then began to assist Ribicoff's subcommittee, sharing his extensive collection of safety research data. This body of work was soon turned into the best-selling book, *Unsafe at Any Speed*, published towards the end of 1965.

In the middle of March 1965, testimony on auto safety got underway in the subcommittee and dragged on through the spring with little action and little attention from the media. It appeared to be a recycling of previous inquiries Congress had made over the last ten years. But in July, the mood shifted as the presidents of the automobile companies were invited to testify. At first, Ribicoff maintained a certain politeness and decorum but was not making any progress with the leaders from Detroit. At Nader's prompting, the questions became tougher and more pointed as to the lack of passenger safety in cars. The GM executives argued before the subcommittee that their cars were safe, expressing safety in terms of "reliability." When it was revealed that, industrywide, 20 percent of all cars on the road since 1960 had been quietly recalled to fix faulty brakes, steering and suspension systems, the hearings took on a more sinister tone. The audience and media sat up and took notice. Other senators began to attend, including Robert Kennedy, who asked harder questions and received damning answers. Kennedy pressed

James M. Roche, the president of General Motors, and its chairman, Frederic G. Donner, to admit that during 1964 GM generated $1.7 billion in profits but spent *only* $1.2 million on safety research. That was the turning point in the subcommittee's investigation. When it was subsequently revealed that only twenty-three *cents* per car was being spent industrywide on safety, Detroit had been unmasked. In full view of the country, from the morning newspaper to the evening news, the automobile industry exposed itself as greedy, inept and unconcerned with its own products or customers' safety.

In the following months, the buzzword across the country was "auto-safety." The revelations in the congressional hearings propelled Ralph Nader and his book to national prominence. General Motors' response to Mr. Nader's indictment of unsafe vehicle design was to begin an unsavory investigation into his private life. Worried that he was trying to benefit financially from the 100 or more lawsuits filed against GM for the Corvair, the corporation hired several private investigators to pry into his personal affairs. When one of them bungled the job and was caught in the act, he confessed and GM was publicly forced to apologize to Mr. Nader for what it had done. The secret investigation turned up nothing except that Nader was an honest and forthright individual. This information would cost GM $425,000 in an out-of-court settlement with Mr. Nader.

This act, setting out to discredit one, lone citizen who had criticized a gigantic corporation, outraged the entire country and guaranteed swift action against GM and the rest of the industry in the form of legislation to regulate the safety of motor vehicles sold in the United States. The National Traffic and Motor Vehicle Safety Act was signed into law by President Johnson on September 9, 1966. Out of this legislation grew the National Highway Traffic Safety Administration (NHTSA) and the power to mandate Federal Motor Vehicle Safety Standards (FMVSS). Given a one-year lead-time, the industry was forced to modify its cars to meet seventeen new safety standards by January 1, 1968. These standards ranged from an energy-absorbing steering column, seatbelts for all occupants (with separate shoulder strap for front-seat driver and passenger), to a dual-circuit brake system and interior padding. The industry's response to the new regulations was to whine and drag its feet.

Henry Ford II summed up Detroit's attitude by declaring, "Many of the temporary standards are unreasonable, arbitrary and technically unfeasible . . . if we can't meet them when they are published, we'll have to close down. If we have to close down some production lines because they don't meet the standards, we're in for real trouble in this country." Mr. Ford came out of that generation of American workers and business leaders who had been seasoned by the realities of World War II—who had met that challenge by building fighter planes like the marvelous North American P-51 Mustang (the namesake of the popular new Ford car) in just 117 *days* from drawing board to first prototype. Now, instead of responding to the battle cry for better and safer cars in a timely fashion, this same generation of men who controlled the American automobile industry wallowed in their own indignation, arrogance and ineptitude.

The auto industry, however, was not alone in its contradictory attitude about safety. The government and the American people were also at fault. They were indignant about the automobile being "unsafe," but unwilling to look at their own excesses as a part of the problem. During the Senate hearings on safety, the senators concentrated on how "crashworthy" an automobile was. They failed to set any performance standards to improve the terrible handling and braking of the average American car. Ralph Nader had owned only one car in his entire life and no longer drove. For all his knowledge of postcrash protection, he was unschooled in the dynamics of safe handling that an automobile should have to avoid an emergency situation on the road (as, it appears, were most auto executives). The senators holding the hearings brought with them their own preconceived notions about what constituted "safe" use of an automobile. Senator Ribicoff was obsessed with speed as the sole cause of accidents. Further, he was convinced that a direct link existed between the automobile companies' support of racing teams and the increase in traffic deaths on the road. Robert Kennedy's attitude represented that of many Americans. *Time* magazine columnist Nick Thimmesch recalled a ride with Senator Kennedy back from the subcommittee hearings to McLean, Virginia. With Bobby at the wheel of his Lincoln Continental convertible, they streaked down the Washington beltway at 75 mph, as their safety belts lay unused on the seat next to them. For all his flamboyance and apparent lack of knowledge about proper car design, Henry Ford II provided perhaps the most astute observation: "Americans like to blast along over interstate highways at eighty miles an hour in big cars with every kind of power attachment, windows up, air conditioning on, radio going, one finger on the wheel. That's what they want, and that's what they buy, and that's what we manufacture. We build the best cars we can to meet the taste of the American people." True, the average American car buyer seemed to show little interest in or knowledge of the technical workings of the car or its safety on the road. But instead of trying to enlighten or inform the driving public by providing superior products, Detroit ruthlessly exploited the consumer with an ever changing facade of annual restyling over an unsafe, technologically stagnant running gear—all in the name of perpetuating profits.

This was a cockeyed time in America's understanding of itself. Now, the car and those building it were the enemy while America indulged itself in all that the car could provide. Only 10 to 15 percent were taking advantage of the new seatbelts installed in their cars, but growing numbers were using the plentiful horsepower under the hood of automobiles that were only marginally safer than before. The country was also heeding the call to "See the U.S.A. in a Chevrolet"—the Sunday-night mantra during commercial breaks on *Bonanza*. With the interstate system half complete and the number of people killed on the road topping 50,000 for the first time in 1966, all highways were being overextended at ever higher speed, given road and vehicle design. The climbing numbers on the speedometer were enjoyed in conversation and actual use, while the *other* number was being denied or furtively bandied about in fleeting talk about those unlucky, poor unfortunates who must have done something wrong to deserve their fate.

As the next few years passed, posted and unofficial speed limits across the country grew further apart. The federal government recorded that average speeds on the rural interstate had reached 65 mph in the early '70s, with 85th percentile speeds at 73 mph. Out on the road, however, among the state troopers who lived in the real world, the picture appeared quite different. One Minnesota trooper remembered his third day on the job back on July 3, 1970. This was the eve of the big holiday for highway fatalities and his own baptism into speed. Working twenty-five miles south of Duluth, Minnesota on I-35, this state trooper clocked twenty-five cars in a two-hour period, seven of which were given speeding tickets. The *slowest* one was going 102 mph! We had moved into the realm of not knowing what we were doing anymore. We were sucking every moment of driving pleasure out of the car while, more often than not, receiving a wink-and-a-nudge from the law of the land in an unofficial capacity. But don't get caught.

In the waning days of 1973 on my momentous trip to Montana, we crossed the entire state of South Dakota on I-90 at over 90 mph, slowing down only to stop for gas. Even though the interstate was posted at 75 mph in those days, there was little or no attempt at enforcement—the unwritten rule of the freeway as long as you didn't blatantly sail by the state patrol. On the return trip, we cruised once again along the wide-open Bad Lands at a relaxed 95 mph. At some time during the trip, I watched a car slowly gain on us through my side-view mirror. First a speck on the horizon, it pulled closer within a

Rural Freeway
SPEED LIMITS
1973

STATE	SPEED LIMIT
ALABAMA	70 DAY, 60 NIGHT
ALASKA	70 MPH
ARIZONA	75 MPH
ARKANSAS	75 MPH
CALIFORNIA	70 MPH
COLORADO	70 MPH
CONNECTICUT	60 MPH
DELAWARE	60 MPH
FLORIDA	70 MPH
GEORGIA	70 DAY, 65 NIGHT
HAWAII	65 MPH
IDAHO	70 MPH
ILLINOIS	70 MPH
INDIANA	70 MPH
IOWA	75 DAY, 65 NIGHT
KANSAS	75 DAY, 70 NIGHT
KENTUCKY	70 MPH
LOUISIANA	70 MPH
MAINE	70 MPH
MARYLAND	70 MPH
MASSACHUSETTS	65 MPH
MICHIGAN	70 MPH
MINNESOTA	70 MPH
MISSISSIPPI	70 MPH
MISSOURI	70 MPH
MONTANA	R & P DAY, 65 NIGHT
NEBRASKA	75 MPH
NEVADA	R & P
NEW HAMPSHIRE	70 MPH
NEW JERSEY	60 MPH
NEW MEXICO	70 MPH
NEW YORK	65 MPH
NORTH CAROLINA	70 MPH
NORTH DAKOTA	75 DAY, 65 NIGHT
OHIO	70 MPH
OKLAHOMA	70 MPH
OREGON	70 MPH
PENNSYLVANIA	65 MPH
RHODE ISLAND	60 MPH
SOUTH CAROLINA	70 DAY, 65 NIGHT
SOUTH DAKOTA	75 DAY, 70 NIGHT
TENNESSEE	75 MPH
TEXAS	70 MPH
UTAH	70 MPH
VERMONT	65 MPH
VIRGINIA	70 MPH
WASHINGTON	70 MPH
WEST VIRGINIA	70 MPH
WISCONSIN	70 DAY, 60 NIGHT
WYOMING	75 MPH

few minutes to reveal a carload of six adults. They briefly drew even with us, crossed over in front and moved over into the right lane. In a few more minutes, they were, again, a black speck on the other horizon. This was driving for many Americans in the early 1970s. Manifest Destiny extended to a freeway system that now crisscrossed the entire country, reaffirming a belief in ourselves, the automobile and driving—without accountability. It did not seem to matter or cause need for concern that from 1940 to 1973, a total of 1,320,447 Americans lost their lives for this high price of automotive glory.

The absolute certainty of our convictions, however, was about to come to a screeching halt.

CHAPTER 2

National Maximum Disgrace

On October 6, 1973, Egypt and Syria invaded Israeli-held territories. Yet within a matter of days, Israel had once again decisively repelled the invaders. In a blitz of shuttle diplomacy, U.S. Secretary of State Henry Kissinger, traveling between Moscow, Cairo and Tel Aviv, shored up an uneasy truce, but the damage was done. The Arab world had suffered another humiliating defeat at the hands of the Jews. As a matter of course, America sided with its long-time ally, Israel, and in response to that support, the Arabs used their last and most effective weapon to punish the United States and the rest of the western world: an economic embargo shutting off the flow of oil. In a matter of weeks, the 1.2 million barrels of imported oil shipped each day into this country trickled down to nothing. America was in shock. The very foundation of its prosperity and way of life was cheap, imported oil. The mighty giant had been temporarily brought to its knees by an attack on its Achilles' heel. To a government and nation already in turmoil over the unfolding Watergate scandal, preventive measures taken to combat the embargo were piecemeal at best. Homeowners and businesses were encouraged to lower thermostats to 68 degrees. Daylight-saving time was extended year-round. And in an effort to quickly reduce gasoline use, a national 55 mph speed limit was imposed across the country.

America was long since used to the free speed of the open road, so only the perceived severity of the crisis forced a majority of drivers to slow down. The sight of endless lines of cars and trucks wrapped around city blocks, waiting for hours to fill up with now precious gas, sent a strong psychological message that slowing down was serious business. Television reporters stood in the middle of empty freeways, showing the entire country the embargo's impact. Government officials, law enforcement and common folks began encouraging each other to slow down and do their part. Even on the *Tonight Show*, Johnny Carson told the nation his drive to work was only a few minutes longer, holding the line at 55. The studio audience neither laughed nor applauded. They simply sat grudgingly in stunned silence.

Our free and easy lifestyle had crumbled before our eyes. The magnitude of the crisis slowed down even the most ardent speed demon. Almost overnight, speeds above 75 mph vanished from the American highway. Grumblings from truckers and over-the-road salespeople about longer running times that moved fewer goods and services were met with deaf ears. Fuel prices were also climbing faster than freight rates, a double hit for the trucking industry. Little comfort came from the wording of the Emergency Highway Energy Conservation Act of 1973, which stated that the 55 mph National Maximum Speed Limit was a *temporary* fuel conservation measure. When the Arab oil embargo was finally lifted in March 1974, things slowly began to return to normal. There was talk, but little action, to return speed limits back to precrisis levels. But by year's end, a new development had come to light, one that would cause the national speed limit to be made permanent in January 1975. During that year's time, the number of people killed on the road had dropped from 54,052 to 45,196—an amazing decline of 8,856 people. Not since World War II had there been a greater one-year drop in the number of Americans killed on the road. When it was further revealed that the number of miles driven had not dropped in corresponding fashion, the National Maximum Speed Limit was declared an important new lifesaving program for U.S. highways.

Now, the minority of "speed kills" fanatics within the government, insurance industry and law enforcement had their golden opportunity. They had been decrying the evils of speed for years and now had their "proof," being able to back any foe into the wall with numbers so great that no one could dispute them. This was also good news to the burgeoning movement of Ralph Nader's "consumer advocates" interested in improved automobile crash protection. Wielding ever more power in Washington, "Nader's Raiders" based their safety emphasis on saving lives after a collision had already occurred, a philosophy that was speed critical. The slower a vehicle was moving, the better chance there was to save lives after a crash. Further, controlling safety, whether it be speed limits or auto safety, was a much easier job when directed from Washington, D.C. rather than at the state level.

A powerful guilt factor worked its way through the American people. Every driver knew how he or she had driven before 55. The sheer freedom and happy-go-lucky spirit of those days gave way to a more somber attitude that they had somehow aided and abetted the profligate use of speed, wasting precious gas and causing the loss of thousands of lives. Now was the time of penance. No longer should an American think of driving as fun or enjoyable. Complicit in this guilt was the highway patrol. It had long looked the other way when most highballers sailed on by. These troopers, more so than any other Americans on the road, felt responsible for letting far too many "speeders" go, reasoning that if they had stopped more of them, some might still be alive today. This feeling would cause most state police agencies to hit the road with a vengeance against any speeder. No one was going to fly down the road and kill again.

The inducement to force each state to comply with the reduced speed limit was that no federal highway funds would be doled out to any state with a limit higher than the national 55 mph maximum. This mandate was strict and absolute. Any state raising its speed limit or failing to be certified for enforcing the limit would immediately lose all of its federal highway funds. Each year, a state would have to be recertified for its performance in enforcing the new limit. Federal speed monitoring stations were set up at various points within each state to ensure that speeds were accurately monitored. A compliance system would record the quality and quantity of a state's progress in slowing its citizens down—and keeping them slowed down. This new way of doing business helped to create an avalanche of paperwork and bureaucracy, shuffling compliance reports, speed data and the number of 55 mph citations from each state to the federal government.

Everything appeared to be working rather well for the first two years of 55's existence. Average speeds on the rural freeway system dropped from 65 mph in 1973 to 58 mph from 1974 through 1976. Fatalities hovered around the 45,000 mark, and drivers reluctantly, but with a certain diplomatic aplomb, received millions of speeding tickets which were only too cheerfully handed out by troopers who were now convinced that this was the only way to save lives. With new federal funding, state patrols kicked into high gear with a whole battery of enforcement techniques. Aircraft increasingly patrolled the major highways of most states, and unmarked police vehicles of almost endless variety began covertly nailing violators of the national speed limit. Some agencies used old cars and trucks to hide radar guns inside, sometimes parked as stalled vehicles on the shoulder. These unconventional practices garnered plenty of speeding citations, but drivers quickly caught on, treating every parked car on the road as suspect. Troopers began to dive more frequently through the center median of the interstate to nab highballers going the other way. This was one of the more risky endeavors by the state patrol. Once, while I was driving through Illinois in the mid-1970s, I watched as an overzealous state trooper forced a retired couple completely off the road into the ditch as he dove through the center median to nab a speeder. He never returned to see if they were all right. No one was going to get in his way of running down a dangerous speed demon. As with any war, technology was developed to meet the needs of police agencies trying to slow down an entire nation. Improvements in radar quickly developed from a stationary to a mobile mode. No longer would a trooper have to work in tandem with a parked patrol car and its radar unit. Each trooper was now freer to hunt on his, or increasingly, her own, extending coverage over a longer length of road with squad cars so equipped.

This dramatic crackdown on "speeding" brought a shift in attitude among both the driving public and law enforcement. Within the ranks of state police agencies, a whole new breed of "ticket writing machines" was being recruited. Trained in the latest techniques of speed enforcement to the exclusion of all else, these aggressive young rookies became the new "True Believers" in the national speed limit. The fact that virtually every driver on the road was fudging a few miles per hour on 55 meant that these new troopers were looking at *everyone* as a lawbreaker worthy of a speeding ticket. The training of their older counterparts in the three "Cs" of enforcement: Courtesy, Courtesy and Courtesy, gave way to a less respectful, more authoritarian demeanor that treated every driver as the enemy. This harsher attitude helped to create a growing mistrust between drivers and law enforcement. The average motorist began looking over his or her shoulder more often, on edge that the law was out to get them. Drivers began to internalize their own "speeding" as justifiable, reasoning that it was just a small amount over the limit, not enough to affect overall safety.

With the oil crisis over, the perceived need to slow down to save gasoline was gone. Slowing down for safety, at the personal level, was an invisible issue. All but the most timid drivers could convince themselves they were not going to have an accident and, therefore, did not have the time to drive exactly 55. Over the years, speeds began to creep back up as most motorists found the conscious effort to hold speeds down too mentally fatiguing, unless their cars had cruise control. Even then, the rate at which the road passed by seemed so fitfully slow, most could not keep it under 65 mph. Troopers, on the other hand, viewed this as a dangerous trend and spent ever more time writing ever more speeding tickets. These new speed enforcement fanatics began racking up hundreds, in some cases, thousands of tickets. It did not help public relations matters when police agencies of many states, such as Wisconsin, began taking only credit cards to pay for speeding fines. Many drivers resented hearing, "No cash, no checks, Visa or Mastercard only." This did little to create a feeling of camaraderie between drivers and the law, smacking more of raising revenue than the level of safety on the road.

Drivers, too, were beginning to look at each other as the problem. Individually, one could rationalize the need to speed, but seeing another driver zoom past was cause for outrage. *Here* was the speeder who was the threat to safety. Most motorists never realized that the same was being thought about them when they were passing someone else. This development brought a new level of hypocrisy to the American road. "Everyone else is the problem, not me," was becoming the mindset in the mid- to late 1970s. A new kind of self-righteous motorist began appearing in greater numbers on the highways of the country. Convinced it was their duty to slow down other drivers, these motorists moved from the right, slow lane of traffic into the left, passing lane of the highway. These "Left-Lane Bandits" started to cause some major traffic tie-ups, especially during rush hour when a handful of these slowpokes could block all lanes of traffic, causing serious congestion. In the days before 55, slower drivers would keep to the right and allow faster vehicles to slip past uneventfully. This unspoken courtesy of the road made for improved traffic flow at higher speeds. Now, the New Think for 55 looked upon a fast, smooth flow of traffic as a bad thing. Some states stopped posting minimum speed limits on the freeway, and removed Slower-Traffic-Keep-Right signs. More and more drivers began floating from lane to lane, impeding the flow of traffic and forcing other motorists to slow down. Law enforcement and safety experts hailed this "improvement." These "good drivers" were blocking the progress of speeders. What they failed to consider was the growing trend of faster drivers weaving between slower ones, and the increased level of aggravation caused by these new, artificially created traffic jams.

The last half of the 1970s was a curious time in terms of America's attitude towards driving. The automobile had become the preeminent form of transportation in the U.S., but was now looked upon by many as a bad thing. The pollution, congestion, death and suffering—which were significant, at times—mounted on the negative side of the ledger, aided by the Ralph Nader school of thought. Within the government, a strange mixture of good and bad legislation, regulation and social tinkering had a profound effect on America's outlook on itself and the car. Increased governmental regulation was forcing a recalcitrant American auto industry to conform to the dictates of a small minority of auto safety advocates in and outside the government and insurance industry. The country was fighting itself ever more vociferously in a battle to bring safer cars to the road with lower pollution and higher gas mileage. The rules and regulations being handed down by the National Highway Traffic Safety Administration and the Environmental Protection Agency were not only being challenged by the American car manufacturers as too costly or ineffective, but also by the safety and clean-air movement as too lenient or too slow to be implemented. Watchdog groups like Nader's Center for Auto Safety and the Insurance Institute for Highway Safety (through companies like State Farm Insurance, which the Institute represented), were spearheading endless litigation for more safety which inspired reverse lawsuits from the American auto industry.

This push-comes-to-shove struggle was setting up an environment for the foreign car industry to invade the American market. Both European and Japanese manufacturers were positioning themselves to take advantage of the adversarial relationship America was having with itself by quietly meeting the various mandates for safety, increased fuel mileage and lower pollution. To its credit, the Nader movement had helped to bring rudimentary safety improvements to cars, along with lower pollution emissions. But the oil embargo magnified the need for better fuel economy for the average American car, so the government set up an aggressive schedule to raise the level of miles per gallon from an average of 14 mpg in the mid-70s to 27.5 mpg by the mid-1980s. Gasoline prices had long since been higher in Europe and Japan, so this was one area where the foreign competition was already prepared. A smaller, more fuel-efficient engine was also easier to clean up in terms of air pollution—another dividend.

This influx of foreign cars almost instantly satisfied an American car market hungry to get more miles per gallon. To a nation previously monopolized by the domestic automobile industry, this wave of imported cars allowed average Americans to compare—perhaps for the first time—what they had been driving and what was available on the world market. Many were stunned to learn how bad American cars really were, having been lulled into complacency by a soggy ride with vague steering and spongy brakes.

I remember driving a friend's brand-new Audi four-door sedan back in 1974. I was amazed at how precise the steering and suspension were. Scooting around corners and changing lanes were crisp and sharp. The brakes slowed the car down like hitting a bank of sand. Though the engine did not have the brute force of an American automobile, the Audi achieved brisk acceleration by revving the four-cylinder motor through a manual, 4-speed gearbox which was quick to shift and well-matched to the horsepower of the car. A rate of 30 mpg was possible on the highway at 60 mph. After an extended period driving the Audi, I was perplexed why we here in America were not striving to build cars of such response and quality.

More and more Americans were also making this discovery, and the number of foreign cars imported into the country steadily increased as the antagonism between government, auto industry and consumer advocates grew. The shortsightedness of all sides on the national issue of automobile transportation did nothing to inspire confidence in the American people. The industry was looking to short-term solutions, hoping the onslaught of regulation would be temporary so big cars and bigger profits would come back into favor, leaving them free to continue with business as usual. Senators, Congressional representatives and safety advocates took an increasingly dim view of the perceived robber-baron capitalists in Detroit. During this time, such antagonism helped to produce some of the worst cars ever built in America: oversized, overweight and underpowered from add-on emissions equipment.

By 1977, this tug of war with regulation reached its apex, evolving from useful to oppressive and extreme. Certainly, bad cars had come before bad-car laws, and there was no doubt that things like 3-point seatbelts and the Corporate Average Fuel Economy (CAFE) standard *were* helping to save lives and clean up the air. These new automobiles were marginally safer, more fuel efficient and substantially better in terms of lower air pollution with the introduction of unleaded fuel and the catalytic converter in 1974. But the mindset in Washington seemed to be: If a little activism worked small miracles, then a whole lot more must yield even better results. Into this alphabet soup of advocate, bureaucrat and car manufacturer, a disciple of Ralph Nader's, Joan Claybrook, ascended to the throne of the National Highway Traffic Safety Administration (NHTSA) under President Jimmy Carter. "Saint Joan," as her detractors called her, went on a relentless campaign to rid America of all the death and suffering caused by the automobile. Government would now reinvent the car, dictating to Detroit how "safe" vehicles must be built.

To be fair, this pattern of adversarial mandates had started long before Claybrook's takeover as Administrator of NHTSA. Detroit still had not gotten the message on safe vehicle design, but increasingly, the regulations handed down by the federal government were less than innovative. At a time when reducing weight on a car meant extra fuel economy, heavy, bulky 5 mph bumpers were being required front and rear. Questions of freedom and governmental intrusion also reached a pinnacle. When the auto industry was forced to introduce the famed ignition interlock in 1974 (preventing the engine from starting unless the front seatbelts were engaged), a nation of new-car buyers was so outraged, the mandate was repealed by a special act of Congress. With Claybrook at the helm, this tradition of antagonism by government continued unabated. Several "safety" cars were built under contract for NHTSA to show Detroit the strides that could be made with "crashproof" automobiles. And some did achieve outstanding results for protecting passengers in accidents as high as 50 mph. The downside was the tremendous expense of development passed onto the taxpayer, and styling that was "interesting" to say the least. Some of these vehicles were downright ugly. The aesthetic appeal of form and function—the potential beauty the automobile could possess—had given way to crash protection at all costs. The Claybrook attitude was, if people were dumb enough to get behind the wheel in the first place, they should at least be afforded the security of driving a Sherman tank.

At this time, a renewed interest in "passive restraint" took shape in the form of the airbag and automatic seatbelts. Fewer than 20

"Research Safety Vehicles" developed by the federal government. Long on crash protection, short on looks, these automobiles cost the taxpayers millions, but brought little change to the products coming from Detroit. U.S. DOT

percent of Americans were buckling up, so some way had to be devised to force them to wear a belt or to cushion the blow from striking the interior of the car in an accident. The airbag had patents dating back to the 1950s, but was not sufficiently refined to be practical until the early 70s. Even then, the cost per unit was quite high, anywhere from $500 to $1,500. The device itself was an ingenious array of sensors, wiring and a nylon bag built into the front end and steering wheel of the vehicle. In a crash, the sensors would ignite a canister of sodium azide, creating a gas that filled the bag, preventing the driver's head and upper body from hitting the wheel. Installation was not just limited to the driver's side, either. Any passenger could have airbag protection if properly designed. Passive seatbelts, on the other hand, were not nearly so unobtrusive. Attached directly to the door, the shoulder strap forced the driver into contortions just to sit down in the seat, the belt pulling and chafing at every turn. Future innovations promised a more civil, motorized version that would wind its way up the door jam after an individual was seated. Both of these devices had serious drawbacks at the time, requiring further development. The airbag saved lives only in front-end collisions, and then, only if seatbelts were used. It was also subsequently learned that short people, or small children in safety seats, could be severely injured and, in some cases, killed when sitting too close to the exploding bag. As for the passive belts, they were quite often so uncomfortable that people simply disconnected them. Their one-size-fits-all nature didn't properly fit many drivers or passengers, either. Motorized shoulder straps without a lap belt were the worst of all, actually causing severe neck injuries in many cases when motorists were hung up on the strap as they slid forward in a crash. Both airbags and seatbelts did have potential for saving lives, but not in these crude forms. To Claybrook, however, they were just what was needed to save people from themselves, and she pressed on—unsuccessfully—during her years as Administrator to get them into cars. At the time, many people expressed fears about a bag that could burst open while they were behind the wheel. Also, the short-term retooling of the steering wheel and dashboard would have been an excessive burden on an already struggling domestic auto industry.

While this battle was raging on, a further development in safety and speed would help to revitalize Detroit, but drive a wedge between Claybrook and the safety community itself. Because of the need for increased fuel economy, the automobile was decreasing in size and weight. In safety terms, this meant less room for crash protection. Many safety advocates, long since used to the adage "Bigger is Always Safer," were alarmed by this. Claybrook, though, was not disturbed by the trend. She felt that a smaller fleet of vehicles—all cars being about the same size on the road—would maintain crash safety. The safety community began squabbling amongst themselves about what was best. Most of them failed to realize that this downsizing was producing another dividend in safety they were unprepared for. Smaller, lighter-weight cars were more nimble. They stopped quicker, turned tighter, and in general, handled better to *avoid* accidents in the first place. Another by-product of this was that more motorists were taking advantage of this improved design by driving faster. These performance safety improvements were now making greater strides than passive crash protection. Much to Claybrook's chagrin, the area of safety she had given the least attention was helping to maintain some parity in the car-size versus lives-saved battle. Her response to this "dangerous" trend of increased speed in these smaller, better designed cars was talk (and little else) of mandatory speed governors in vehicles. She was joined in the fight by safety pioneer William Haddon, president of the Insurance Institute for Highway Safety and long a fierce believer in a mechanical speed limit to save lives. But there was no political will for such a device, so Claybrook had to settle for a mandate permitting only an 85 mph speedometer in new cars and trucks. Her thinking was that it made it "look" like you were driving faster than you really were. She never figured this dim-bulb idea had a different reality for many motorists, who would drive that much harder to bury the needle to the peg.

Since this attack on car design, and more specifically, the American auto industry, was not yielding enough lifesaving results, Joan Claybrook trained her sights on the American people themselves as the problem. There was the growing trend of rising speeds on the nation's highways, aided by a growing backlash against federal speed-limit control. Most states, especially in the West, resented the federal government's intrusion into their domain. Bumper stickers began appearing in states like Montana stating, "55 Be Damned!" The speed limit was baptized the "Double Nickel" by truckers and "Nixon's Revenge" in the automotive press. In response to such public antagonism, state legislators began introducing 55 mph repeal bills. Some missed being passed by a handful of votes; others were passed only to be vetoed or left unsigned by the governor when the federal government threatened the direst of consequences if a state repealed the national limit outright. State politicians began looking for loopholes to skirt the law. Although the states had to certify their enforcement of the speed limit, they were not responsible for the type of penalties assessed to the driver. States could make an end-run around 55 by decriminalizing speeding to varying degrees. Nebraska began charging only $10 for violations up to 65 mph on the interstate. Minnesota followed suit with a similar law for any 55 mph road. And Montana was by far the easiest of them all, charging only $5 for any violation of speed, no matter how high, during daytime-driving hours. More importantly, these new laws did not help insurance companies jack up their rates. Up until this time, most states recorded points for speeding on a driver's record, and premiums were arbitrarily raised on millions of motorists receiving these tickets. This was a lucrative, profit-making surcharge for the insurance industry that did nothing to improve safety on the road. Now, with this loophole in 55, states could still draw revenue from speeding tickets but drivers were less inclined to gripe because their insurance companies were none the wiser. It was not long before most western states had only a minimal fine for exceeding 55.

Also to Claybrook's dismay, technology had become available to the average driver to combat the increased use of radar for speed enforcement. These radar detectors, or "fuzzbusters" as they were affectionately called by one manufacturer and many irate motorists, worked with varying degrees of efficiency at picking up the radar energy coming from the police unit. Some of these devices were so poorly designed, they barely sensed the signal until a trooper was almost in direct beam with the radar gun. Others were extremely sensitive, picking up the faint signal reflected off of other cars or signs over hills and around curves. Faster drivers quickly armed themselves with these more expensive units and created a whole new battleground between the police and motoring public. In the late 1970s, before many state patrols recognized just how powerful some of these radar detectors really were, many troopers left their radar guns running continuously wherever they drove, creating a golden opportunity for these detector-equipped drivers to covertly slip past, slowing down long enough to miss being detected, then speeding up on their way.

Even with the loophole laws, radar detectors and, also, CB radios for motorists to inform one another about speed traps, Claybrook decided to tighten the screws further on speed enforcement to make these wayward states and the American people conform to her dictates. Her solution to these problems was to mandate, in 1978, a new set of federal compliance criteria for 55. This would permit only 70 percent of motorists to exceed the national speed limit in 1979, 60 percent by 1980 and so on to 1983, when no more than 30 percent would be allowed to exceed 55. If a state failed to meet these goals, up to 10 percent of its highway funds would be taken away. But money and incentives would also be made available to states that fell in line with increased enforcement.

Pressure was also being applied within the safety community itself to unify support for 55. There was a substantial body of evidence about the safety and effectiveness of 85th percentile speed limits that the federal government had amassed before the national speed limit took effect. The traffic engineers who had pioneered these studies were now being displaced by the new wave of Slower is Better. There was no new evidence, as yet, to dispute these studies aside from the reality of reduced deaths on the road. But those engineers who had done the previous research were now having their arms twisted to recant and "believe" in the national limit. The landmark studies by David Solomon, with follow up work on the interstate freeway system by Julie Cirillo and others, had now fallen out of favor. Their considerable work in the area of speed, speed limits and their relationship to safety didn't fit the New Order of 55. Now, treated as pariahs in the field of highway safety, they were relegated to the back room or left the government entirely. Nothing was going to get in the way of total commitment to 55. It did not prevent some of them from continuing further research, now shunned by the safety community that spawned them, and denounced by the new faction within that movement that deified 55 above all else.

While these machinations were going on behind the scenes, the federal government was also quietly changing how it monitored and recorded speeds across the country. Previously, speeds had been measured on straight and level roadways with at least a four-second separation between vehicles. This recorded the highest free speeds of cars and trucks unimpeded by anything. Now, the new system took into account congestion and all manner of road anomalies, including speeds on hills, curves, even in front of stop signs and signal lights. Simply by juggling statistics, speed trends could be lowered anywhere from 3 to 6 mph, depending on conditions and time of day. This helped to make it "look" like the whole Claybrook speed-enforcement plan was working. And this was further manipulated by other loopholes in the compliance system, allowing states to "adjust" their speed data by factoring in speedometer, sampling and measurement "errors." By averaging in all the speeds in a state's road system posted at 55, the numbers could be further fixed to fit the requirements of the federal government.

With all of these excesses, the result was predictable: a corrupt, draconian system that favored bureaucratic double talk and outright falsification of data which was justified for safety's sake. Claybrook's fanatical devotion and belief in crash protection, reduced speeds and arcane mandates to enforce them only succeeded in turning the nation against itself. But the tide had turned on this sort of activist government. It was the last gasp of New Deal liberalism started by Franklin Roosevelt in the 1930s, raised to new heights by President Kennedy and Johnson in the 60s, and it was running headlong into big trouble with Carter by the late 1970s. Activism in government had translated to more bureaucracy and problems for our economy rather than meaningful change that was of benefit to society. This was aptly apparent in "Saint Joan's" running of NHTSA. And the kind of backward-thinking mentality that exemplified the

Claybrook era ultimately brought its reward to the American highway: Speeds, fatalities and even the death rate rose during her tenure, leaving behind a legacy of mismanagement, overzealous regulation and decline in economic productivity and social spirit. The very "malaise" President Carter spoke of in a speech to the nation was something his administration had helped to foster.

Change was brewing as the 1980 presidential campaign got underway. For one thing, the Republican platform called for the outright repeal of the 55 mph National Maximum Speed Limit, saying that the most cost-effective program was for each state to set its own limits. In the deregulatory environment created by President Reagan after his election, he did not rush right out to repeal 55. There was still a formidable coalition in and outside the government that called for the speed limit to be retained. The political consequences of repealing the speed limit and suffering the fallout of increased fatalities were reinforced by "experts" in the field of traffic safety. Within the Congress, there was also a powerful block of senators and representatives—mostly in eastern states where lower speed limits had been a lifelong reality—who held tight against the prospect of ending federal speed-limit control. At the head of this group was Representative James Howard from New Jersey, who chaired the House Public Works and Transportation Committee. Howard was "the father of the 55 mph speed limit," and had worked with President Nixon back in 1973, suggesting a compromise between a 50 mph speed limit for cars and a 60 mph limit for heavy trucks. He had the most to lose if 55 was voted out as a failed public safety policy. Long entrenched in Washington, Howard held incredible power in terms of pork-barrel politics that funded road transportation projects around the country. His absolute devotion to *his* 55 mph speed limit meant no repeal would come soon, if ever, as long as he was calling the shots.

This did not mean that the nation's trend to further weaken the national speed limit would not continue. The federal government would have to deal with the growing reality of more drivers exceeding 55. It had become quickly apparent that Claybrook's aggressive compliance program would never be met. And by 1981, it would be increasingly difficult to meet the 50 percent requirement. Anything lower would be impossible. If the more stringent requirements *were* enforced, the growing frustration with the speed limit would likely boil over into outright repeal. This was something the safety lobby was unprepared for, and politicians looking towards reelection were not too thrilled with the possibility of a potential blood bath manifesting itself on the highways. As a result, the Omnibus Reconciliation Act of 1981 changed the compliance criteria to remain at 50 percent while also cutting back federal funding for enforcement—the by-product of the Reagan administration's desire to cut domestic spending in favor of a beefed-up military budget. Now, only $20 million would be spent on extra 55 enforcement, with fewer dollars doled out to hungry state patrol budgets, as state legislatures had to cut or reprioritize spending to handle the increased burden of governing with fewer handouts from Washington. The outcome of this was also predictable. The epidemic of speeding went on for the most part unchecked, except for the occasional enforcement blitz that nabbed a handful of hapless victims. This decline in enforcement and the public's lack of cooperation only furthered resentment on both sides. Troopers were seeing a growing problem, increasingly out of control, with declining budgets and, in some cases, reduced personnel to deal with it. This created a morale problem within the force, which was sworn to uphold a law no one on the road was obeying. Drivers were also becoming more adept at beating the system. Many armed themselves with an array of improved, superheterodyne radar detectors, some with radar-absorbing bras for the front end of their cars. A few outlaws even had illegal radar jammers to disrupt the signal coming from the radar gun. The battle was escalating and the law was losing.

Two grassroots movements were to have a further impact on changing America's attitude about the national speed limit. The first got its start on a spring afternoon in 1980 as a thirteen-year-old girl walked along a street in Fair Oaks, California to a church festival. Struck from behind, she was killed by twice-convicted drunk driver, Clarence Busch, out on $100 bail after two days in jail for his third DWI arrest. The young girl's mother, Candy Lightner, channeled her grief, outrage and anger into the organization Mothers Against Drunk Driving. The MADD movement quickly spread across the nation, with several other offshoots: SADD (Students Against Drunk Driving), AIM (Alliance against Intoxicated Motorists) and RID (Remove Intoxicated Drivers). The statistic that half of the fatal accidents in the country were due to drunk driving galvanized the movement, which tried to crack down on the problem by getting laws changed from state to state. This brought a change in the national attitude as to why fatal accidents were occurring on the road. Law enforcement attention was diverted to address this new outrage over drunk driving.

Now, state patrols would have to split their time between the growing need for speed enforcement and the increasing demand to do something about the drunk-driving problem. Many troopers, while still firm believers in the need for reduced speed, had long been frustrated by the reoccurring problem of pulling mangled bodies from wrecks caused by drunks behind the wheel. It was a difficult position for law enforcement to be in. Speeding violations were relatively easy to write up. Using a radar gun and ticket book, a trooper could write several in the span of one hour. This helped to generate substantial income to perpetuate the whole routine: more tickets, more income. Drunk driving was a time-consuming affair of field-sobriety testing, blood or breath analyses, final booking and jail. There was no shortage of either kind of driver, especially on the weekends. But most troopers grasped the reality of the situation. They knew who was really causing the more deadly problem out on the road. Over time, dealing more exclusively with drunk drivers would divert attention from 55 mph enforcement, giving the unspoken go-ahead for drivers to speed up even further.

The other grassroots organization was founded in January 1982, in Dane, Wisconsin. In the heart of the dairy belt, the Citizens' Coalition for Rational Traffic Laws finally created a formal and vocal opposition to the 55 mph speed limit. The president of the CCRTL, James J. Baxter, was an affable Everyman with an easygoing smile and demeanor, devoid of the usual dogma associated with those who supported the limit. His opposition to 55 stemmed from the pragmatic observation: How can this be working if no one is obeying it? The nation didn't exactly rush out in droves to join Mr. Baxter's organization. Eight years of 55 propaganda and more than knee-jerk support by almost every business, safety organization and government office had taken its toll. Further, the motoring public was now driving down the road with a level of hypocrisy that was difficult to fight. National polls continued to show all but unswerving support for the speed limit by the public at large. Eighty percent favored retaining the limit even though increasing numbers were admitting to violating the same limit they supported. This attitude was working its way through the American psyche with potentially dangerous results: It is all right when *I* speed, just as long as everyone else doesn't. This was a seemingly insurmountable challenge for Baxter and his fledgling Coalition to overcome, and at the outset, an unusual cadre of eccentric speed demons quickly joined up. But Baxter persevered, bringing into the fold an impressive array of automotive journalists, doctors, and even traffic engineers for his advisory board. Out of this more solid foundation, a respectable membership of automotive enthusiasts slowly but steadily began joining the rank and file of the CCRTL. Soon, state chapter coordinators began popping up all over the country and the stage was set for a more formal opposition to 55.

The initial response to the Coalition was usually indifference, or even outright hostility. This was the more visceral, emotional reaction to decades of death and suffering on the highway that never seemed to end. After all, people believed the motto, "Speed Kills" and there seemed nowhere else to go from that point forward. Most people simply bought into the "saves-gas, saves-lives" jingle that seemed to appear endlessly on billboards, in telephone-book inserts and radio spots. Few seemed willing to question why they believed in 55, even though most were not strictly adhering to it. It seemed the right thing to do. However, this early opposition to 55 by the Coalition did help to generate a growing controversy in local newspapers across the country, touching off a diatribe in the letters-from-readers column. It was emotional, flamboyant, not too technically accurate, but America was ever so fitfully beginning to take a look at why it had a 55 mph speed limit.

This same question was also being asked more pointedly in the halls of Congress. Several legislators, mostly from western states, were beginning to form their own coalition within government to start a drive to repeal the limit. Many states were running dangerously close to noncompliance, even with the new regulations, and the thought of losing federal highway funds was moving some politicians to action. Congressional supporters of the limit were also taking notice of falling compliance and diminishing support. To justify tighter sanctions and the need for greater enforcement, Representative Howard decided to fund a study by the National Academy of Sciences to investigate the "benefits" of the 55 mph speed limit. From the start, this study was described as supportive of the limit, designed to show why it was a good thing to retain. Some researchers within the Academy wanted greater freedom to look more dispassionately at the limit, hoping to achieve a sounder scientific interpretation. Howard would have none of it. This study would justify why the limit should be retained, not look at both sides of the issue. However, the small number of anti-55 senators and representatives wanted "costs" as well as benefits to be explored, and after intense debate, Howard finally relented. This meant the panel overseeing the study would require some members whose support for 55 was not total and absolute.

Separate from this Congressional challenge to the study, other organizations were trying to influence its outcome. Jim Baxter and his Coalition were trying to be a part of the panel, along with the now-former NHTSA secretary, Joan Claybrook, and her Public Citizen advocate group—another Nader offshoot. Neither was successful at gaining a seat on the board, but both could give testimony and sit in on panel meetings. The final advisory staff consisted of long-time supporters of the speed limit, but the subject of costs would also be included. Released just after the 1984 elections (to prevent any political controversy being generated), "55: A Decade of Experience" reiterated that "the 55 mph speed limit has been one of the most effective highway safety policies ever adopted." The report sang the praises of a federal program that still saved 2,000-4,000 lives per year—in spite of falling compliance. This affirmation of the speed limit's lifesaving potential didn't succeed in garnering any new support for more stringent measures to enforce it. The study had told the True Believers what they already knew. And since it so blatantly favored the national limit, it only helped to galvanize support against it. On the *MacNeil/Lehrer Newshour,* the chairman of the advisory committee for the study, Alan Altshuler, tried to defend the study shortly after its release to the public. Jim Baxter of the CCRTL was also asked to appear on the program, and he pointed out the many contradictions and biases the study had. The opposition to the limit had found a firm foothold and was receiving more opportunities to voice its views, with the media only too happy to feed the controversy. Now, every time someone declared that "55 Saves Lives" there was someone not far away forcing them to defend the truth of the statement. The 55 mph speed limit was in trouble. Government support was flagging along with public compliance, and the study failed to create a stricter environment for enforcement. Nevertheless, most Americans were in favor of keeping it, just as long as they could disobey it.

The growing lack of compliance with the speed limit and this hypocritical support of it created an environment where no one was willing to do anything about it. There were now clear lines of division for and against the limit, but the majority of Americans cared little about seeing it changed. Many states now had some kind of bill that got around the national limit, usually for 10 mph over 55. This put de facto speed limits nearer to where they were before federal speed-limit control, so there was little incentive to work to repeal it outright. The door was opened for a slow, steady increase in speed that states were unable, or unwilling, to stop. The federal government was providing no extra money for enforcement beyond the $20 million required by law. As long as each state was able to jockey its speed data using "speedometer error" and show that it was writing the required number of speeding tickets—whether they were worth the paper they were written on or not—each state received its highway funds, and all was well.

But the inevitable started to happen. Some states long since used to cooking the books on speed data were not able to cheat any longer. Even with speedometer error they couldn't attain the minimum 50 percent compliance. By 1985, Maryland, Vermont and Arizona had failed to meet the federal standards. If it weren't for the fudge factor in the monitoring system, almost every state in the nation would have been out of compliance and threatened with the loss of funding. The federal government then scheduled hearings for the three states in trouble. Within the confines of these hearings, an impressive battery of enforcement techniques was explained to an eager-to-please federal staff. No one actually wanted to take away states' highway funds—the very money designed to keep the roads safe in the first place.

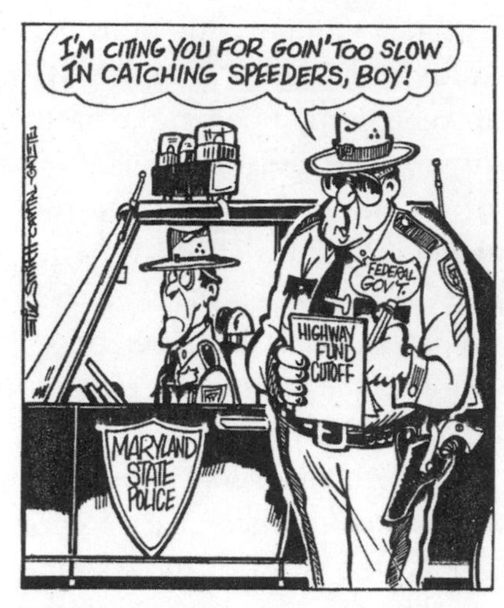

Typical political cartoon of the day. Big Brother in less than sinister repose. Eric Smith - The Capital Gazette

Enforcement by each state police agency was novel to say the least. Using unmarked cars, parked vehicles and even troopers dressed as hitchhikers with battery-operated radar guns in knapsacks, law enforcement took to heart the task of trying to force an entire state to slow down. Airplanes and special speed task forces with sinister names like B.A.T. patrol and "Wolfpacks" hit the road. One program even tried to get kids to spy on their parents to tell them to slow down—all in vain. To the relief of each state in question, they were put on a special probation to see what else could be done to get people to drive 55. The most innovative of these new measures was the "rolling roadblock," driving with squad cars abreast at exactly 55 mph to create a gigantic caravan of vehicles trailing behind just under the speed limit. This caused extreme aggravation to the impeded motorists stuck behind this snail's-paced traffic jam. A handful even tried to take the shoulder to beat the system, which, of course, was futile. One other technique was more

Maryland, 1985. "Rolling Roadblock" to force motorists to drive 55 mph in an attempt to save the state's federal highway funds. Jim Powers - Maryland DOT

covert, and actually illegal: Having troopers sit right in front of the speed monitoring devices to make sure every vehicle passing it was counted at exactly the speed limit. The federal government was willing to look the other way in an effort to help these wayward states get back into compliance. It worked. Each state was magically recertified in record time to prevent any loss of funds. However, Arizona was fined a token half a million dollars to set an example to any other state close to being out of compliance—money that was later quietly returned for its "cooperation."

The federal government was less than helpful if a state wanted to repeal the national limit outright. In the same year, Nevada had had enough, passing a bill that the governor signed raising the speed limit to 70 mph. The state was willing to call the federal government's bluff, challenging it to a battle in court if highway funds were revoked. For one brief moment, at a small ceremony along a stretch of I-80, the speed limit was raised back to 70 mph. In that instant, the feds cut off Nevada's funding, and the court challenge was on. Returning the limit to 55, Nevada got its money back, but threatened to take the case all the way to the Supreme Court.

Instead, the state let a lower appellate court judgment stand in favor of the federal government. The nation was stuck. 55 was going to stick until an act of Congress returned to each state the right to set its own speed limits.

During this time, a small but growing number of covert, extremely fast drivers started to appear on the highways of America. Long since convinced that there would be no relief from the 55 mph speed limit, they gave up hope and took to the open road armed with the best radar detectors and CB radios that money could buy. They were also behind the wheel of hot cars that had no trouble being safely driven, in the right hands, at over twice the national speed limit. Many of these automobiles were coming from Europe and Japan, but Detroit was also on a comeback, producing ever better cars that had no trouble traveling well over 55. The curious fact of foreign competition had shaken the Motor City out of its lethargic period of poor car design more than any federal mandate. Listening to the strains of Sammy Hagar singing *"Can't Drive 55,"* these outlaw drivers took to the highway with a vengeance to make up for all the lost time the national speed limit had cost them.

I was one of these covert drivers. Behind the wheel of a red-and-black Thunderbird Turbo Coupe, I took to the road with a flagrant disregard of the national speed limit, having known all too well what it was like during the good old days before 55. I was a fast driver, but not a foolish one. I flew when the road was open and chugged along when the cops and traffic were present. I defeated everything state patrols could throw at me. Radar, unmarked cars, airplanes, you name it. I was blessed with the gift of thinking like a cop, realizing where would be a good place to hide a speed trap, and usually being right. I was also incredibly lucky. I missed being caught by mere seconds in some cases. My trusty radar detector helped in many instances, its Geiger counterlike pulses and lights informing me which type of radar was in the area: X, K-band or instant-on. It became second nature to look over my shoulder when passing under any overpass to see if a trooper was sitting on the entrance ramp. After a while, the radar detector became more of an annoyance than an aid, I usually saw nine out of ten patrol cars before they saw me. It made for terribly stressful driving. I spent more time looking for the law than watching the road. I never had a close call with an accident, but always being on edge made driving more aggravating than enjoyable. In the mid-80s, most state police agencies had still not caught on to this type of covert driving, though some began equipping troopers with faster Mustangs and Camaros to equalize the fight. In reality, covert, fast drivers represented a tiny minority on the road. But other motorists liked to tag along behind these illegal highballers to make some easy time, letting the faster guy ahead take the fall in most cases.

A buddy and I, on a trip from Minneapolis, Minnesota to Ithaca, New York, never slowed below 80 mph except when we stopped for gas or slowed down for cops. Taking this trip from the center of the country to the New England states, I discovered a secret that other drivers had perhaps learned. The Canadian route from Detroit to Niagara Falls dramatically cuts travel time, and not so much due to the shorter distance. It works because the Canadians don't travel under 80 mph on their highway system, even though radar detectors are illegal and the speed limit is 100 km/h (63 mph). On the return trip, we ran the gauntlet of New York, Pennsylvania, Ohio and Indiana state police, meeting up with two young women from Marengo, Illinois who *passed* us. Driving in tandem, we crossed the state of Indiana in record time, covering the 160-mile distance from Ohio to the Illinois border in under two hours. The rest of the trip was just as uneventful, and we shaved off several hours on a journey that would have taken much longer had we adhered to the 55 mph speed limit.

The influence of this small band of high-speed drivers beating the system began to filter its way through the automotive-enthusiast press into the general driving population. That you could beat the speed trap and get away with it only helped to add to the overall feeling of lawlessness. It also reinforced the feeling that *I* know what I'm doing behind the wheel, while everyone else is an idiot who should still drive 55. This outlook, along with a continued denouncement of speed from the safety community, helped to drive an even bigger wedge in the country's feeling about the limit. This them-against-us mentality was entering into every facet of the problem with no relief in sight. Neither of the factions, pro and con, was really attracting any new members. America was only too happy to continue speeding with its radar detectors, loopholes in the law and hypocritical attitude that this was the way it had to be. No one was providing any alternative solutions that would reconcile the increasing desire to speed and the need for greater safety. Groups like the AAA, trucking industry, insurance and safety associations weren't going to budge an inch in the direction of higher speed limits. The American Motorcycle Association now joined with the Citizens' Coalition for Rational Traffic Laws against the speed limit, but most car buffs complacently chose to drive down the road with their radar detectors and CBs instead of taking the high ground and joining Baxter's Coalition. Since the organization did not take any definitive stand on safety issues like seatbelts or drunk driving, it became increasingly difficult to rally support. The Coalition only declared that each state would have to work out its own safety agenda after repeal. No one was willing to work together to deal with the tangled mess the public safety policy had become. Only the growing problem of noncompliance—the public voting with its foot instead of the ballot box—was moving the issue to a head.

By 1986, the problem of speed had reached critical mass and action had to be taken either to repeal the limit or revise the compliance requirements once again to allow a greater percentage of speeders to exceed 55. The Reagan Administration went on record favoring a return to states' rights on the speed-limit question. Supporters of the limit were still looking for some way to increase enforcement and force people to slow down, in order to give the national speed limit more credibility. But they were losing ground, and the entire weight of the issue was beginning to give way in favor of repeal. Then, a potential compromise presented itself. The National Academy of Science's study, "55: A Decade of Experience," had opened the door to the possibility of raising the speed limit on rural stretches of the interstate freeway system to 65 mph in an attempt to preserve 55 over most of the nation's highways. The interstate carried 20 percent of the country's traffic, which would help to put every state back into compliance. Debate was fierce and pointed. Many supporters wanted to hold out to the bitter end to preserve what they thought to be the most valuable safety program on the American highway. Others were more confident that maintaining 55 for most highways, with the 65 mph rural freeway exemption, would help to prevent its total repeal. Those against the national speed limit now rallied with renewed energy and increased members. The possibility of raising the speed limit, even on the interstate, made most auto enthusiasts and truckers sit up and take notice.

Attached to a larger transportation bill filled with plenty of pork-barrel spending, the 65 mph rider and the bill itself were vetoed by President Reagan as too wasteful, but Congress overrode the veto by a one-vote margin. Most states quickly moved to raise their speed limits on the rural interstate, taking advantage of not only the higher speed, but the freeway exemption from the federal speed compliance program. Within a year, thirty-eight states had raised their limits to 65 mph, and several others had legislation pending. The opposition to 55 hailed this as a victory and a positive first step toward total repeal of national speed-limit control. Truckers, over-the-road business people and just plain ordinary folks seemed to find the higher limit a bit more comfortable on long trips. Safety advocates and most state patrols watched uneasily. Some had tried to make suggestions to Congress. One was Brian O'Neill, now filling William Haddon's shoes as president of the Insurance Institute for Highway Safety. He recommended things like a national seatbelt law and outlawing radar detectors to offset the possibility of increased fatalities. But so many years of uncooperative, divisive talk between those for and against the limit had left its mark. The new limit would stand without any extra safety precautions. The average American would be the guinea pig as 33,000 miles of rural interstate would become a testing ground to see what effects the higher limit would have. Virtually everyone on the freeway was a willing participant, however. Public opinion still showed large support for 55, but now a majority appeared who favored the freeway exemption. Troopers, while skeptical, liked not having to write endless tickets, because most drivers were holding it pretty close to the new limit. The True Believers in 55 were less convinced. O'Neill of the Insurance Institute predicted that drivers wouldn't obey 65, either. The stage was set to see what the outcome would be after one year's time.

By the end of 1987, well over three-quarters of the states had adopted the new 65 mph limit for their rural interstates. The hope was that this 10 mph increase would decriminalize the speed the driving public was traveling at and solve the majority of the speeding problems on our road system. The interstate was the country's safest road by design. This small increase seemed to be harmless, keeping the limit under the already established speed parameters of the freeway. Another less talked about factor was the exemption from the federal government's compliance system. States raising their limits did not have to report speed trends on that portion of their rural interstates now posted at 65 mph. This helped them to meet the 55 mph mandate for all other, slower roads. The handful of eastern states that chose to retain 55 on all their roads immediately cried foul. They would be punished for keeping 55 in force on their rural freeways, because it would still be a part of the monitoring system. To that end, Congress worked out a moratorium to suspend *any* compliance penalties, but still required each state to monitor and submit its speed data each year. Now, no state could be punished for being out of compliance until the federal government could reevaluate the compliance system and make changes to ensure monitoring would be fair for all states involved.

The year passed and the number of people killed on the interstate system was tallied. About 500 more people had been killed on the freeway network than one year before—very close to the estimates that were given by many experts. Some of this increase was attributable to shifting traffic patterns, more motorists switching from neighboring secondary roads posted at 55 to the faster rural freeway. But the raw number was undeniable. Whatever the cause, fatalities had gone up. However, this increase in death and suffering made only momentary headlines in newspapers around the country, quickly fading from the watchful eye of the press and public at large. No one wanted to be reminded about what was suspected might happen. This higher death toll didn't bring a return of the 55 mph speed limit to the freeway. On the contrary, 65 was deemed an acceptable trade-off between getting to your destination quicker, or not getting there at all. More importantly, the new limit, despite higher fatalities, seemed to ensure a 55 mph speed limit for all other highways. If there was this dramatic an increase on our safest road—a 10 mph increase and 500 additional fatalities—then raising the limit on any other highway was doomed because the risk would be far too great on a lesser road.

For most Americans, the speed-limit issue began to hang in limbo. No one except the most fanatical supporter of the national limit wanted to go back to 55 on the rural interstate, and no one but the most ardent speed demon wanted to pursue total repeal. The motoring public continued cruising on down the road, ignoring both extremes. As long as drivers could get away with exceeding 55 by 10 to 15 mph, no one wanted to do anything about a situation that seemed to have no real solution.

Safety emphasis began to shift away from speed and the lifesaving potential of 55. The 65 mph limit had helped to do that. It became increasingly hard to chime in with "55 Saves Lives" when all but seven states in the Union had gone to 65. Even the safety movement began to shy away from direct references to 55, diverting attention to drunk driving and to demanding that cars have airbags. The public had grown tired of someone constantly telling them to slow down. Cracking down on drunk driving had much more appeal, whether or not the police had any real success in dealing with the problem. Besides, drunk driving did not attack the majority of motorists on the road the way slowdown campaigns did. Most states started to look elsewhere, trying "voluntary" seatbelt laws and roadside sobriety checkpoints to reduce the annual death statistics, with only marginal success.

By the early 1990s, a new development began to revive the speed-limit debate behind the scenes. It appeared that Brian O'Neill's prediction was coming true. A growing percentage of motorists was beginning to exceed 65 by significant margins. At various test sites around the country, the number of cars over the limit had jumped from 5 percent just after the limit was raised in 1987, to 35 percent exceeding 70 mph by early 1993. More importantly, it appeared that in states that still retained 55, the percentage exceeding 70 was not nearly so great. Part of this increase was due to the federal compliance system. States with the 65 mph limit could tell the truth about speed increases because their rural interstates were no longer a part of the monitoring program. The 55 mph states were still fixing their speed statistics to make it look like more motorists were obeying the law. But, with the raw data in hand, those in favor of keeping America slowed down quietly prepared to make a final stand on behalf of the National Maximum Speed Limit. The federal government's new compliance requirements were on the verge of being implemented in 1993—a tangle of regulations which prorated roads by design in relation to the flow speeds of traffic and penalized faster roads by greater margins than slower ones. The new requirements were now so confusing, no one understood them. They had also been revised many times to ensure that most states would comply at the outset, permitting the "speeding" that was already going on. And the new regulations still left a couple of small loopholes that would unfairly penalize the handful of 55 mph states.

Slowly, the system was moving forward to deal with the federally mandated speed limit again—something states really hadn't had to do since 1987. Instead of a rally cry for a crackdown on the speed limit, the reverse was beginning to happen. Bills to repeal the speed limit quietly started to make their way through Congress. Backed by Jim Baxter's Coalition (now renamed the National Mortorists Association), the Fair Speed Limit Act of 1993 called for setting speed limits between the 67th and 85th percentile speed on any road, effectively repealing the national limit. Long-time supporters of 55 also began calling for substantial revisions or outright repeal. The AAA was recommending a 65 mph limit on the entire interstate system, both rural and urban. And in April of 1993, the American Association of State Highway and Transportation Officials (AASHTO) ended almost two decades of 55 mph support, calling for each state to have the right to set its own speed limits. Only the diehard supporters of 55, who were rapidly losing followers, were really interested in sticking it out. Fewer politicians were looking at the prospect of repeal as the political suicide it was once thought to be. With the Republican takeover of the U.S. House and Senate in

November 1994, Congressional consensus was building to return the issue back to the states and let them worry about it.

Which is exactly what happened.

Through the spring and summer of 1995, an amendment to abolish the 55 mph speed limit (with sponsorship arranged by the National Motorists Association) steadily made its way through House and Senate committee hearings. Becoming part of the National Highway System Act of 1995, the bill sailed through the Senate on November 17 by an impressive 80 to 16 margin. The next day, the House passed it on a voice vote. During the following weeks, the White House was lobbied heavily by safety advocates to veto the legislation. Ralph Nader tried in vain to see President Clinton personally to tell him that *6,400* more Americans would die on the road if the bill were passed. However, the Highway Act also included funding for $6.5 *billion* in state road programs, so it came as no real surprise that the bill was signed into law. With the stroke of a pen, the federal 55 mph National Maximum Speed Limit ceased to exist.

Western states wasted no time raising speed limits. Many of them had had automatic repeal bills on the books for years, waiting for the day when 55 would go the way of the dinosaur. After December 8, the first day the new law took effect, Texas went to 70 mph, Arizona and Nevada to 75 mph. In typical Big Sky fashion, Montana thumbed its nose at the federal government, returning to "reasonable and prudent"—no clearly defined speed limit—for daytime driving hours. By the summer of 1996, half the states in the Union had raised their rural freeway speed limits to 70 or 75 mph. The speed of traffic picked up too. The 85th percentile speed rose from 72 mph in December '95 to 78 mph by April '96 on Nevada's 75 mph interstate. Interestingly, on Montana's freeways the same increase was observed, even though much higher speeds were permitted—as witnessed by a friend who (with his heart in his throat) successfully passed a state trooper at 95 mph on I-94 in March of that year. And much to the chagrin of the Montana Highway Patrol, when a trooper did write up a speeding violation, a small but determined band of highballers were routinely beating these 100+ mph citations in court, convincing jury after jury that they were driving in a "reasonable and prudent" fashion. Most Americans, however, were enjoying their new found freedom without putting the pedal-to-the-metal— *yet.*

Dunagin's People: *By Ralph Dunagin*

"The 55-mph speed limit was last observed on this spot."

© Tribune Media Services. All Rights Reserved.`Reprinted with permission.

Throughout 1996, the national media, federal government and safety organizations watched discreetly for the expected blood bath on America's highways. Newspapers like the *Denver Post* repeatedly called the state patrols in Wyoming, Montana and elsewhere around the country for accounts of gory high-speed pileups—finally giving up when the body count numbered no more than the year before. Behind the scenes, after the raw statistics were tallied up and released by the U.S. Department of Transportation in March 1997, the True Believers in the whole *55 Saves Lives* campaign sat in mute disbelief. The number of people killed on the road remained virtually the same. In 1995 (the last year of 55), 41,798 road fatalities were recorded. The following year, 41,907 died—just 109 more—a statistical wash. This revelation went all but unreported by the national media at the time. Safety groups also chose silence about this dramatic turn of events, concentrating instead on controlling the damage to their prestige over the airbag fiasco. For two decades they had battled with the auto industry and federal government to have them installed in cars, but the bags themselves had succeeded in reducing the annual death toll by only 1 or 2 percent at most, not the 33 percent promised by safety engineers. They were also killing dozens of small adults and children, a negative tally on the safety ledger. The press was all too willing to hector safety advocates on the growing number of decapitated kids due to airbag deployments in low-speed collisions, but never once demanded an explanation as to why speed didn't kill this time around. In the years following, the media would give passing attention to stories like Montana's Supreme Court throwing out the state's vague "reasonable and prudent" speed limit in December, 1998 or to recent safety studies that "proved" more people were really killed when the 55 mph speed limit was repealed. But through it all and leading up to today, organizations like the National Safety Council, Center for Auto Safety, and the Insurance Institute for Highway Safety had become impotent. Their two most important lifesaving programs, airbags and the 55 mph speed limit, were of little or no value. They had cost Ralph Nader's beloved consumers *billions* of dollars in higher vehicle prices and *billions* of hours in lost travel time. It is a tragic and disgraceful legacy. More than 40,000 people are still dying on the road each year, and this same safety movement is incapable of doing anything about it. Apathy now rules the American highway. Vision, action and cooperation are at an all-time low.

There is a way out, however.

On the other side of the Atlantic, the nation of Germany has undertaken an aggressive lifesaving program that does not come at the expense of speed. Over the last quarter century, Germany has reduced the number of people killed on its roadways by almost *70 percent* even though half of its *Autobahn* freeway network has *no* speed limit. Impressive results can be achieved when you stop

Rural Freeway
SPEED LIMITS
1999

STATE	SPEED LIMIT
ALABAMA	70 MPH
ALASKA	65 MPH
ARIZONA	75 MPH
ARKANSAS	70 MPH
CALIFORNIA	70 MPH
COLORADO	75 MPH
CONNECTICUT	65 MPH
DELAWARE	65 MPH
FLORIDA	70 MPH
GEORGIA	70 MPH
HAWAII	55 MPH
IDAHO	75 MPH
ILLINOIS	65 MPH
INDIANA	65 MPH
IOWA	65 MPH
KANSAS	70 MPH
KENTUCKY	65 MPH
LOUISIANA	70 MPH
MAINE	65 MPH
MARYLAND	65 MPH
MASSACHUSETTS	65 MPH
MICHIGAN	70 MPH
MINNESOTA	70 MPH
MISSISSIPPI	70 MPH
MISSOURI	70 MPH
MONTANA	75 MPH
NEBRASKA	75 MPH
NEVADA	75 MPH
NEW HAMPSHIRE	65 MPH
NEW JERSEY	65 MPH
NEW MEXICO	75 MPH
NEW YORK	65 MPH
NORTH CAROLINA	70 MPH
NORTH DAKOTA	70 MPH
OHIO	65 MPH
OKLAHOMA	75 MPH
OREGON	65 MPH
PENNSYLVANIA	65 MPH
RHODE ISLAND	65 MPH
SOUTH CAROLINA	65 MPH
SOUTH DAKOTA	75 MPH
TENNESSEE	70 MPH
TEXAS	70 MPH
UTAH	75 MPH
VERMONT	65 MPH
VIRGINIA	65 MPH
WASHINGTON	70 MPH
WEST VIRGINIA	70 MPH
WISCONSIN	65 MPH
WYOMING	75 MPH

treating the average motorist as a lawbreaker, and have a safety movement that asks the driver to participate in its safety agenda.

The whole face of highway safety changes dramatically when you deal openly and honestly with the issue of speed.

CHAPTER 3

State-of-the-Art Driving: The German *Autobahn*

For the American who loves to *drive* and is battle fatigued by the "speed kills" mania which has plagued the United States over the last couple of decades, the German *Autobahn* is like going to automotive heaven. Driving this marvelous system of expressways means leaving behind your radar detector, police paranoia and a small amount of your personal freedom in return for a motoring experience unrivaled anywhere else in the world. To the newcomer, especially one fed a lifelong diet of "55 Saves Lives," the *Autobahn* can be a bit unsettling. The high speed, confusing international road signs, and the abrupt manner in which motorists flash turn signals and headlights to pass can leave the first-time driver nonplused. However, once the handful of rigidly obeyed rules of the road are understood, the German superhighway network transforms itself from apparent life-threatening chaos to a freeway for traveling long distances quickly and safely.

The *Autobahn* is a no nonsense mode of transportation. It is not a place to wander from lane to lane while reading the newspaper, talking on the cell phone or holding the dog in your lap. (These things happen with all too frightening regularity in America.) It is a freeway for attentive *driving*. Concentration is the rule. Drivers are in command of their vehicles, not vice versa. By philosophy and design, this road system takes advantage of speed, using it with great

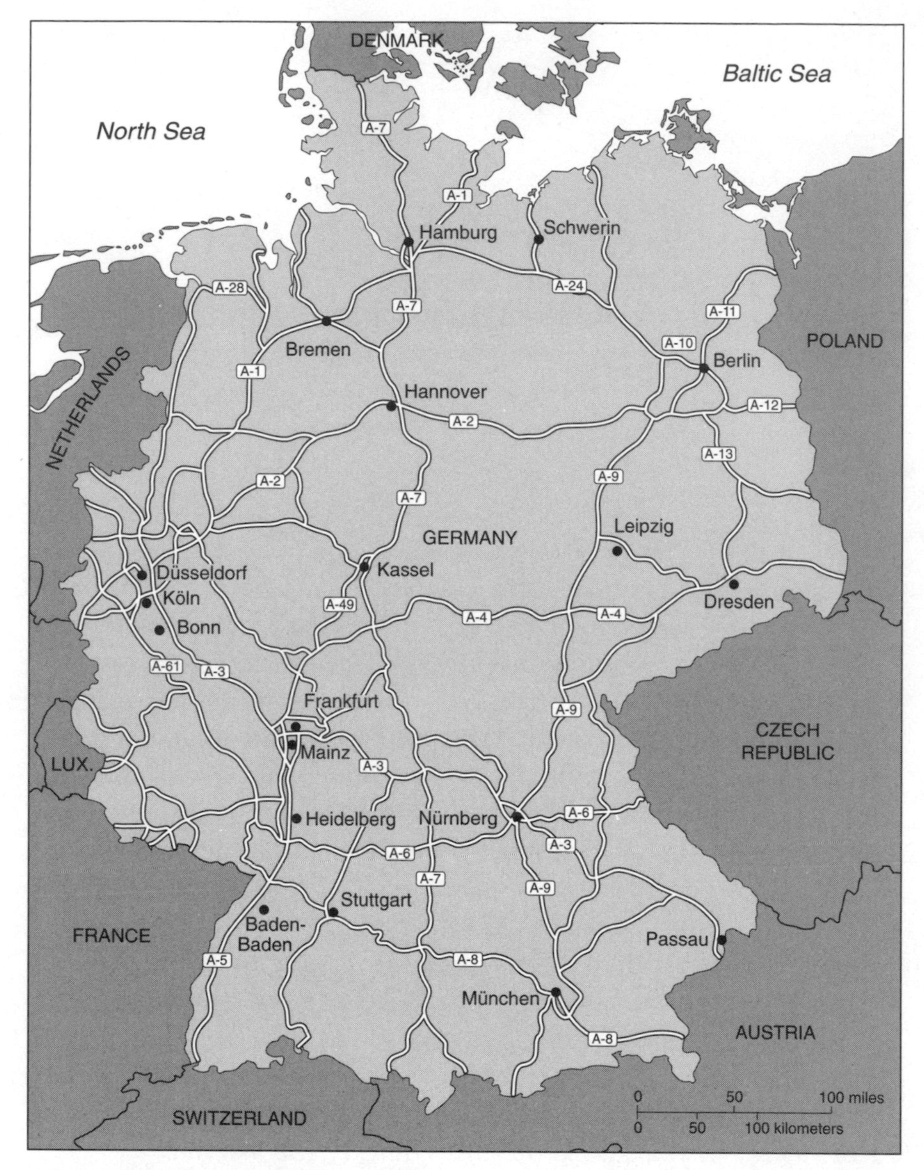

The 7,000-mile Autobahn network. In a country about the size of Montana, half of Germany's freeway system has no speed limit.

efficiency, as does rail and aviation transportation. It is not uncommon to drive 300 miles in under three hours, arriving no more fatigued than if you had covered the same distance in four hours at 75 mph. The only difference on the *Autobahn* is that you are keyed up with the exhilaration from *driving* your car, not weary from a monotonous trip down the American interstate.

My first experience with the *Autobahn* was in June of 1985. On a three-week vacation to cure myself of the 55 mph blues, I found that my rented BMW 316, four-door sedan was exactly what I needed for driving pleasure. The deep, supportive bucket seats wrapped snugly around the driver and passengers, each provided with 3-point seatbelts. The four-wheel disc brakes with antilock control were coupled to a 1.6-liter, four-cylinder engine that moved the car smartly down the road, burning 98 octane premium *leaded* gas. The only drawback, I later found out, was the 4-speed stick shift. No 5th gear...

Puttering down the city streets of Mainz, heading south in the direction of the German freeway, I suddenly realized we were on a surface street that simply flowed directly onto the A-63 *Autobahn*. The first indication of this transition was the international road sign every fast driver quickly learns to identify: the gray circle with diagonal lines through it—the end of speed restrictions. Not since my days in Montana had I enjoyed anything like this, and the BMW was light years ahead in terms of drivability over the old, boat-anchor Chrysler I had fought to keep on the road twelve years earlier. This was as smooth and tight as driving a train on rails. The engine revved cleanly and briskly through the gears, never hesitating or missing a beat, thanks to the high-octane fuel and BMW's attention to

The sign all fast drivers long to see on the Autobahn—the end of speed restrictions. American Autobahn Society Archive

engine balancing. The stark whine of the well-tuned motor settled in at maximum rpms as the speedometer needle rested comfortably near the 180 km/h mark (110 mph). I hadn't driven this fast in a long time and it showed. The sides of the freeway were streaking by in a disconcerting fashion. This feeling of moving very fast put a sharp tingle up my spine. Within a few minutes, though, my brief sense of fear passed and the exhilaration remained. I wasn't sure if it was the speed which set me on edge, or the American fear of getting caught by the police. The road ahead, however, was empty this sunny June morning, and after a minute or two, we came upon our first two cars, one passing the other about a half mile down the road. Quickly closing the gap, familiar as I was with Americans blocking the left passing lane of the interstate, I backed off the gas. The ring of the engine and speed of the car dropped—my foot just touching the brake. The distance was now down to less than a few hundred yards when, wonder of wonders! With plenty of room to spare, the car in front of us promptly pulled over into the right lane to let us pass. Jumping back on the gas, I accelerated smartly by, quickly building our speed back up to 180 km/h. Time after time, the same magical thing happened. A car passing in the left lane would quickly move back into the right to let us through. I rarely, if ever, had to back off as these well-trained drivers, traveling in the range of 80 to 90 mph, moved over at the first opportunity. I felt like I was King of the Road—until it was my turn to promptly move over and let an even faster car pass.

This was paradise. The BMW simply ate up the miles—ah—kilometers. The steering and suspension allowed me to sweep the car through long, drawn-out curves at top speed with no wheel wander, precisely placing it in the center of the lane. After a half hour at this speed, I felt comfortable and relaxed, taking in an occasional glance at the lush, green hillsides that the *Autobahn* cut through with double-time efficiency. But there was one thing missing. Catching up with myself, I popped a cassette into the deck. For most, fulfilling a high-speed fantasy like this would probably cause them to rock to the classic strains of Springsteen's *Born to Run* or a hundred other rock-and-roll road songs. I drive to a different, more timeless beat, adding to the exhilaration of driving over 100 mph by taking

Beethoven's music and Schiller's poetry to a new level they probably never dreamed of: *Laufet, Brüder, eure Bahn. Freudig, wie ein Held zum Siegen!* (Run, brothers, down your road. Joyfully, like a hero to victory.) In their homeland, I had finally achieved what I had long been searching for: A road and a government that permitted me—playing by its rules—to drive as fast as I wanted.

That's kilometers per hour. . . . A legal 110 mph on the A-5 Autobahn near Baden-Baden, Germany. June 1985. Photo by the author

---

The *Autobahn* was the world's first superhighway, predating the Pennsylvania Turnpike by seven years. In fact, the Turnpike's design was based on the layout of the German model. The *Autobahn*/freeway idea began on the drawing board about 1927, but the worldwide depression that hit in late 1929 slowed down its development. The program, however, was quickly co-opted by the Nazi party when it came to power in 1933. In September of that year, the first section opened for travel from Frankfurt to Darmstadt (about twenty miles) with great fanfare and Nazi pageantry. The following year, the Third Reich unveiled a plan for a *Reichsautobahn* network some

6,900 kilometers long (4,300 miles). This state-sponsored project would help to ease the crushing unemployment the depression had brought, stimulating the economy and showcasing to the world the accomplishments of the new German state. By 1936, more than 1,000 kilometers (600 miles) were completed and opened for traffic. As predicted, the new superhighway was a public relations bonanza. The Nazi-controlled government was quick to promote the achievement by taking foreign dignitaries on zeppelin rides over the thin sliver of white concrete cutting its way through the rolling countryside. Those drifting high above marveled at the handful of cars streaking by on the pavement down below—many faster than the airship itself. Right from the start, the road was designated with no speed limit and drivers wasted no time finding out how fast their cars would go. Even Adolf Hitler got into the act, riding along with his chauffeur in a supercharged Mercedes at over 100 mph. (This is one time one wishes that high-speed driving *were* more dangerous!)

It is interesting to note that the German car manufacturers, at the time, resisted the idea of a nationwide system of superhighways with no speed limit, fearing their cars could not sustain speeds of 60 mph or more. Another vexing problem was that the average German did not own a car. Many stretches of *Autobahn* in the mid-1930s saw

The road to a high-speed future. An early stretch of Autobahn near Frankfurt, Germany in the mid-1930s. Heinrich Hoffmann/ LIFE Magazine © Time Inc.

less than four cars per minute—fewer than 5,000 per day. Germany's problem was the reverse of America's. There were roads but no cars. So Hitler contacted one of the leading automotive engineers of the day, Ferdinand Porsche, with his idea for a "People's Car:" a roomy, four-seat vehicle with an air-cooled engine, because most Germans did not have a garage for winter parking and such an engine did not require antifreeze. This motor car was to have a 100 km/h (63 mph) cruise speed and return 30 to 40 mpg, keeping operating cost low. Even before Porsche's audience with *Der Führer*, the engineer had already designed and built several prototypes of a similar layout. Under the direction of the Nazis, several more were built and then paraded before the country. But, before the car could be manufactured, Hitler's megalomania turned away from cars for the people to *Lebensraum* and conquest. Only out of total defeat and ruin would Porsche's *Volkswagen* emerge as a German success story after World War II.

The *Autobahn* itself did not fare well after the War. Many sections were destroyed and slow to be rebuilt, given the more pressing need for food and housing. By 1960, only 1,600 miles were open for traffic. However, the fact that the *Autobahn* and the German auto industry were being reborn together helped to create an environment conducive to modernization and innovation. From the outset, it was clearly understood that there was no speed limit on the *Autobahn* system. It wasn't couched in terms like "reasonable and prudent," nor were there speed-limit signs to be flagrantly ignored as in America. The word *unbeschränkt* (unrestricted) was used to describe the lack of any limit on the German superhighway. To traffic and automotive engineers, this sent a message, in no uncertain terms, that high-speed performance would be critical for both road and car. This fact placed engineer/executives like Ferry Porsche (Ferdinand's son) and the legendary Rudolph Uhlenhaut of Mercedes-Benz at the forefront of vehicle innovation *and* corporate decision-making, while American automotive design stagnated with bean counters running Detroit. The German government also wisely kept gasoline prices high through taxes levied to rebuild the nation's infrastructure—not just roads and bridges, but railroads and other forms of public transportation. This higher gas price guaranteed that automobiles built in Germany would also have to consider fuel economy within the context of the performance necessary to drive on the *Autobahn* at high speed.

Through the 1960s and 70s, the hardship of rebuilding the country began to give way to a growing prosperity that manifested itself in a German *Autowelle* (wave of cars). The average citizen was beginning to look beyond the basic transportation of the *Volkswagen*, using new-found purchasing power to buy more performance from manufacturers like Audi, BMW, Mercedes-Benz and Porsche. The fact that you could cut loose as fast as you wanted on the *Autobahn* promoted the performance aspect of driving. Consequently, more and more Germans were taking to the road with ever faster cars. As in America, this brought with it a dramatic increase in the number of motor-vehicle deaths. In 1960, 14,406 people died on German roadways, a figure which had jumped to 19,193 by 1970—the most people ever killed in a single year in Germany to this day.

The reality of these growing numbers and the already existing emphasis on free speed and performance driving helped to forge very different attitudes towards safety in Germany than in America. The U.S. had long since become accustomed to "speed kills" and "slowdown" campaigns. Cheap gas, increasing horsepower and meaningless speed-limit signs on the side of the road turned speeding into America's "dirty little secret." In Germany, speed was an open, accepted part of the motoring experience. Vehicles and, more importantly, motorists were expected to handle this aggressive and uncompromising driving environment. This attitude of individual responsibility coupled with all-out performance achieved a completely different safety philosophy in Germany than in the United States. By the mid-1970s, as America launched off on a blitz of speeding tickets and efforts to save lives after a crash, Germany began looking to the driver, vehicle and road itself to *avoid* accidents. Since a premium had already been placed on performance because of the *Autobahn's* design with no speed limit, the superior handling of the average German car in terms of tires, brakes, suspension and steering gave it excellent evasive capabilities. This, combined with a comprehensive driver's training program for citizens eighteen years or older who wanted a license, put Germany well out in front to deal more openly with saving lives on the road. This is not to say that crash protection was nonexistent in Germany. Mercedes-Benz had been the leader in the field long before it was known to the average motorist. Since the 1930s, Mercedes had been investigating the dynamics of improved safety after a crash under the direction of passive safety visionary Béla Barényi. However, performance safety was the major emphasis for the German auto industry through the 1950s and 60s. At the time, crash protection was an expensive option being offered by the more upmarket Mercedes-Benz. Demanding that *all* vehicles be revamped for crash protection was a costly and time-consuming affair. That would have to wait a few more years for the cars being imported to the U.S.

This split in safety attitudes between the two countries would not bring immediate results to either nation over the short term. However, better automotive design would permit Germany to weather a period of crisis and prepare it for the long-term goals of economic prosperity and saving lives on the road. When the first Arab oil embargo was imposed in 1973, Germany's long history of high gas prices and more fuel-efficient vehicles not only helped the country to make it through the shock of the shortage, but poised it at-the-ready to export cars to a now fuel-conscious America. Germany, too, faced the oil crisis with a speed limit on the *Autobahn*. From November 24, 1973 to March 14, 1974, a 100 km/h (63 mph) limit was enforced to reduce fuel use. As expected, the number of people killed on the *Autobahn* fell sharply from 1,147 in '73 to 856 in '74 as did the number of miles driven. But, in a nation long since used to the unrestricted speed on the *Autobahn* and a safety program based on performance, the limit was repealed entirely after the end of the embargo. Germany also took the more pragmatic stance that such a slow speed limit was unenforceable and therefore should give way to the historic tradition of no limit whatsoever. As in America, fatalities fluctuated on the *Autobahn* over the next few years, but never again reached their pre-oil crisis levels. On November 22, 1978, the *Richtgeschwindigkeit* of 130 km/h (81 mph) was imposed. This "recommended" speed limit was used to give motorists in passenger cars and on motorcycles an idea of what constituted a reasonable speed for normal driving. Though not enforced as such, it has become the only form of speed-limit control for most sections of the rural *Autobahn* to this day.

A closer look at the specifics of Germany's road safety program is needed to better understand how it comprehensively saves lives and moves traffic. Germany takes to heart the concept of a *road* transportation system. With taxes collected from a $4 price per gallon of gas, they do not skimp on infrastructure. Whether it is the latest stretch of *Autobahn* or a through street in a small town, the foundation of the road itself is such that military tanks can be driven over it. For all of America's talk about national defense over the years, our roads and freeways would not stand up to such abuse. On the German counterpart, this substrate of grading, crushed rock and proper drainage forms the base for a surface pavement that is smooth to drive on and long lasting. Repairs are prompt and detailed—not just a pothole troupe with shovel and pick, but a comprehensive restoration of damaged or worn-out spots that is total and complete. This extends the service life and the time before total restoration and replacement are needed.

A typical stretch of *Autobahn* has two lanes of bituminous pavement for each direction of travel. There is a wide, full shoulder on either side of the driving lanes. The opposing traffic is separated by a short median strip approximately ten to twelve feet wide with a continuous double guardrail in the middle. This almost totally eliminates head-on crashes with vehicles moving in the opposite direction. There are no curbs, just one smooth slab of pavement from the center median to the edge of the shoulder. Guardrails are also placed on the right shoulder where they help to prevent vehicles from running off the road into trees or ditches. Antiglare cones are placed atop the center guardrail to minimize headlight glare at strategic points. Small shrubs are even allowed to grow up between the rails for a more natural, antiglare barrier. It is a true, limited-access design with no at-grade intersections. All roads pass over or under the *Autobahn*, just like the interstate. Shoulders are wide enough to handle parked or emergency vehicles. Reflective delineators are spaced, in general, every fifty meters to show the outline and the

A modern stretch of Autobahn near Passau, Germany. Clearly marked lanes with wide shoulders and extensive use of guardrails promote safe, high-speed driving under no-speed-limit conditions. American Autobahn Society Archive

direction of the road at night. They are low enough to break off and go underneath a car if hit, but high enough to be seen in all but the deepest snows. Game fences and, in some cases, game paths underneath the roadway help to keep road kills to a minimum. Every mile and a quarter, there is a *Notruftelephon* (emergency telephone). Whether for a breakdown or serious accident, these phones bring the required help quickly and safely.

Entrance and exit points are spaced every five to ten miles apart, depending on the terrain. Noise barriers are used as needed—mostly in urban areas. On heavily traveled stretches, signs with programmable faces warn of coming congestion, speed limits or accidents, and even suggest alternate routes. Certain radio frequencies have been set aside to inform motorists as to conditions ahead. About one-half of the system already has some kind of speed limit in force. Some sections use computer-controlled, variable speed limits to help traffic flow more safely at higher speeds with greater traffic capacity.

The use of paint, markings and static signs is thorough and extensive, creating a driving right-of-way that is always clear and easy to see—even in bad weather. On the *Autobahn* and other major highway intersections, the entrance and exit points are marked with heavy, dashed white lines that clearly indicate where the driving and acceleration/deceleration lanes are located. On a long sweeping turn at high speed, there is never any doubt where the driving lane begins and the other lanes end. On a comparable stretch of interstate, it is sometimes difficult or impossible to tell which is which, causing confusion and a possible accident even at modest speeds. The international road signs used in Germany are a new experience for most Americans and can be confusing in the beginning. However, the majority of them are logical and straightforward once a few basic symbols are understood. Their placement as they appear on the road comes in logical order, aiding driving without becoming too intrusive.

Notruftelephon. Emergency call boxes are spaced every two kilometers (1.2 miles) on either side of the Autobahn system. Black arrows on the road delineators (reflective white posts on the shoulder) point in the direction to the nearest one. American Autobahn Society Archive

Maintenance teams cover thirty- to eighty-mile segments of the freeway, keeping things clean. Clearing the aftermath of spilled truckloads or accidents, they also reattach guardrails and signs quickly, so no accidents result from negligence in the *Autobahn's* roadside safety program. Tow trucks prowl the freeway and monitor police radios. They sometimes arrive before the police when motorists call on the emergency roadside telephones. Maintenance is quick, constant and ongoing. No part of the road surface or right-of-way is allowed to fall into disrepair.

As a part of overall road maintenance, all German vehicles must undergo a rigid inspection program every two years (more frequently for heavy trucks). The *Technischer Überwachungs-Verein* (TÜV) has achieved a worldwide reputation for toughness in its inspection of everything from ships to automobiles. While living in

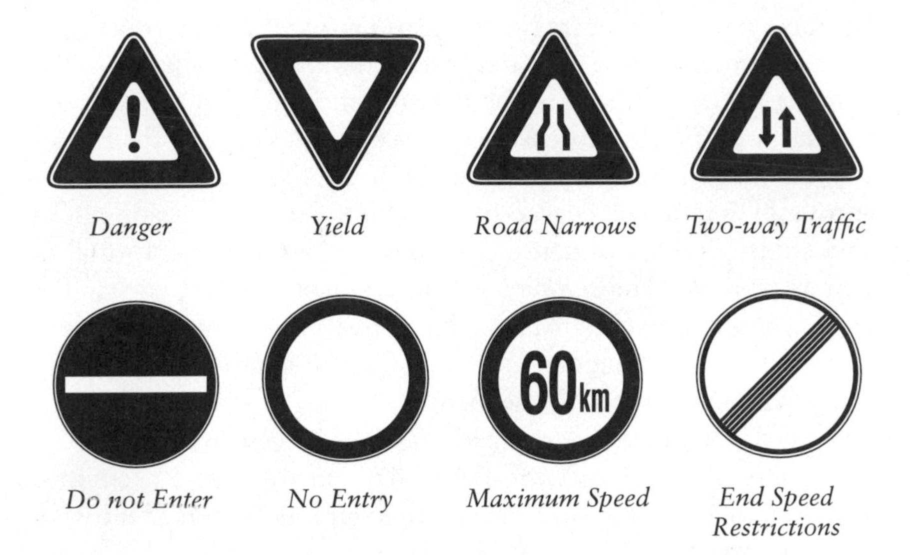

International Road Signs. Confusing at first, but once understood, they are safe and informative.

Germany, I purposely put my Ford Fiesta through the TÜV check-out to see what was required. For the chassis, engine and running gear, a comprehensive testing of emissions, suspension and steering took place. Even the brakes and wheel run-out were inspected, statically "driving" the car on a machine. I was surprised to learn, however, that surface rust, small dings and dents *were* permitted as long as they didn't effect load-bearing areas on the chassis. Once the test was completed, I had a feeling of confidence that the car could be driven at its limits with no guesswork.

As expected, the *Autobahn* has an extensive highway patrol to monitor traffic. Using Audis, BMWs, Mercedes-Benzes and even Porsches, the modern *Autobahn* patrol has squad cars painted white with a green stripe down the middle. These "cucumber sandwiches" patrol the German superhighway with varying degrees of intensity during the year. For example, the APS (*Autobahn Polizei Station*) in Rosenheim on the A-8 *Autobahn* between Munich and Salzburg, Austria is responsible for a thirty-four-mile stretch. On an average day, there may be two cars on the beat. During the peak travel season in summer, a dozen cars and several motorcycles will patrol the same segment. They keep traffic flowing smoothly and safely at the highest speeds possible by ensuring that motorists operate their vehicles responsibly. If a stalled car or accident disrupts the flow, the patrol moves into high gear to bring help, remove vehicles or quickly record the crash, and return the movement of traffic to normal as soon as possible. The summer vacation season is a hectic time for the German highway patrol. As a toll-free expressway, the *Autobahn* carries a tremendous amount of north-south and east-west traffic through Europe. Consequently, congestion can back up for miles even when there are no accidents. The *Allgemeiner Deutscher Automobil-Club* (ADAC, the AAA of Germany) works together with the patrol at these peak times, driving in yellow cars and trucks to aid motorists. These *Gelbe Engel* (yellow angels) are an extra asset of voluntary help, easing the burden the *Autobahn* police have from May through August.

Police Pursuit Porsche! The late Dr. Ferry Porsche donates the one millionth Porsche, a 285 horsepower 911 Carrera, to the Baden-Württemberg police for 140 mph patrols down the Autobahn. Porsche AG

Education of the motorist is taken seriously in Germany. Those who want a driver's license must take a comprehensive training course before they are allowed on the road. Also, they must be at least eighteen years old. This removes a risky subset of sixteen- and seventeen-year-olds who are more accident-prone due to immaturity and poor judgment. Instruction can begin at age seventeen and a half and starts with eighteen hours of classroom training on the theory of driving and the basics of automobile operation. The behind-the-wheel instruction consists of a minimum of eight hours driving with an instructor. Three sessions are on the *Autobahn*, five are cross-country excursions of at least thirty miles per lesson and two training periods are at night, both in and out of town. Trainees are not permitted to drive faster than 100 km/h on the *Autobahn* unless with an instructor. Beyond this basic behind-the-wheel training, another ten to twenty practice sessions are generally required in preparation for the driving test. Besides the obvious skills of parking and turning, recovery from skids and some standard accident-avoidance maneuvers are taught to these new drivers. The ability to think beyond the natural reaction of slamming on the brakes, by steering away from an impending collision, is a valuable part of this basic training course. The German penchant for facts and figures makes the written part of the driving test one of the hardest in the world. Many people fail it on their first try. The actual behind-the-wheel test, while thorough, is less severe. If both are passed, a provisional license is issued for two years, no matter how young or old a driver is. During this time, one must not receive any violations such as driving under the influence, speeding or running a red light. Depending on the seriousness of the infraction, all or part of the training must be repeated. If there are no problems, a full driver's license is automatically conferred after this two-year period. With this third-class license, a person can operate a car or truck up to seven-and-a-half tons. Similar courses must be taken for motorcycles and heavy trucks. More advanced courses are available from the ADAC to improve upon the basic level of skills and proficiency offered by the standard training program. At a cost of $1,500 to $2,000 paid by the individual, driver's training is not cheap, but the German government recognizes one important fact: A well-trained driver is also a safe driver—whatever the speed.

Germany starts early with driver education. In fact, they begin to teach kids the rudimentary rules of the road at a tender age. Within the confines of elementary schooling, *Kinderfahrschule* (children's driving school) puts young ones in pedal cars on playgrounds, using teachers as traffic cops to show kids everything from pedestrian safety to proper attitude behind the wheel of a human-powered car. Besides being fun, this kind of program helps to instill a certain feeling for personal safety and respect for authority that transfers to driving in later life. Drivers with this educational background make wise use of the freedom to choose what is a reasonable and safe speed on the open road.

For drivers already behind the wheel, German television produces a weekly two- to three-minute spot to educate motorists about safety issues on the road. Hardly a program of finger-pointing and dos and don'ts, *Der 7. Sinn* (*The Seventh Sense*) offers real-world safety tips on what to do in an emergency situation from a truck pulling out on you at high speed on the *Autobahn* to lessening the impact of an intersection collision. Staging crashes and other events to show what can happen under extraordinary circumstances, *The Seventh Sense* would make the safety movement in America cringe. This program deals with some of the more unpleasant realities of accidents on the road, and what can be done to avoid or minimize them. These segments treat German drivers as adults, not children. The spot is placed between evening programs during the week on the German version of prime time. Consequently, it has consistently been one of the highest-rated shows on television because of its content and exposure. It is a valuable asset to maintaining and improving driver's training throughout the country.

To offer an anecdotal illustration of the overall effectiveness of this comprehensive education program: The wife of the couple I rented a room from in Germany once had a very close call on the road. On the eve of her seventieth birthday, she was returning home on one of Bavaria's narrow two-lane highways after an overnight visit to a friend. The weather was clear and sunny as she cruised down the road at 100 km/h (63 mph). Coming up to an arrow-straight section, she noticed a large truck approaching in the opposite lane. Suddenly, a red car darted out from behind the truck to pass. In an instant, she drove her little Ford Fiesta right to the edge of the pavement, the right-hand tires on the dirt shoulder. The two cars met dead center alongside of the semi. So close, the red car's side-view mirror was clipped off. They came to a stop on opposite shoulders and a man about fifty-five years old jumped out of the red car and ran over to apologize to my friend. He said he hadn't seen her until it was much too late. After a brief dialog about how no one would have survived the crash if it had happened, they went on their separate ways, shaken but alive.

In terms of safety, this is not the type of incident you would like to see happening in the first place, but training and education saved this woman's life. She completed her German driving course in 1962 at the age of thirty-four. (Most Americans her age have never had *any* formal training behind the wheel.) She did not start driving regularly until 1980, however. Such a close call and its successful outcome point to the need for mandatory driver's training and ongoing education like *The Seventh Sense*. Fortunately, Germany goes to great lengths to keep drivers from becoming a statistic.

Germany takes a dim view of those who drink and drive, especially those who habitually take a serious alcohol problem to the road. The gravity of the situation is taken far more seriously when high-speed driving is commonplace, as it is in Germany. You can't fool around with the possibility of drunks weaving down the road with speeds above 100 mph. This is not to say that this never happens, and some of the more spectacular fatal accidents on the *Autobahn* have occurred when *Geisterfahrer* (drunks who drive the wrong way down the freeway) cause a gruesome crash. Though rare, these crashes prompt the German government and German people to frown on the act of drinking and driving altogether.

They do, however, recognize that alcohol and driving coexist, and they make some accommodation for this reality. When I lived in Germany, the legal limit was .08, just as it is in many American states, but the outcome of driving over this limit is, in general, vastly different in Germany than America. At this blood alcohol level, a person can go out to dinner, have a glass of wine before eating, and another with the meal and still drive home legally. The limit does give the person behind the wheel pause to reflect, and friends don't push booze on the driver, either. The social pressure not to drink and drive is extreme in Germany. This is the single biggest reason for the success they have had in this area. The Germans rate right up there with any nation in consumption of alcohol, but pressure from family and friends is felt if someone tries to get behind the wheel after consuming too much. To be charged with drunk driving is a source of embarrassment and shame. Most Germans picked up for DWI will go to great lengths to keep it from family and friends. Knowing they will be all but disowned through this peer pressure prevents most drivers from imbibing too much.

The laws of the land work together with this societal attitude. Running afoul of the legal system ensures that most drinkers who drive will seriously consider not doing it again. But punishment and fines are on a sliding scale from mild to extreme, depending on the circumstances. What is not in doubt is that punishment is uniform and swift. It doesn't matter if you are a ditch digger or a doctor, the law carries equal weight. For a blood alcohol content as low as .03, a driver can receive a mark on his or her record if pulled over for some other infraction and tested to be "impaired" by alcohol. This mark carries no fine or penalty, but can count against you if other violations occur within the following year, especially another alcohol offense. Such a mark permits drivers to monitor their own experience with alcohol and driving. This run-in with the system also sets up a warning flag to the driver before legal action is taken. Several marks on a driver's record between .03 and .07 can add up to more serious action if this happens during a short period of time.

Between .08 and .11, the first-time offender will receive a $300 fine with points assessed to his or her record and loss of license for one month. The second offense costs $600, points and three months' loss of license; the third, $900, points and three months' loss of license. The loss of driving privileges is absolute. There are no exceptions or work permits, no matter who you are. Friends of our family who live in a small village south of Munich once related to me that the town doctor started to be chauffeured by family and friends. The townspeople learned later that he had been charged with drunk driving, and could only make his rounds in this manner.

For habitual drinkers and drivers, the full weight and measure of the law falls upon those whose judgment is so clouded by alcohol abuse that their need for drink overcomes the fear of punishment. These repeat offenders face mandatory treatment, jail time and loss of license. Depending on the circumstances, once a driver is above .11 blood alcohol concentration, the court may choose to impose a sentence of detox with prison time from one to five years. This is contingent on the number of times a person has been pulled over for alcohol-related driving, or if the person was involved in a serious, life-taking accident. After the time in jail, driving privileges can be taken away for six months to five years. In some extreme cases where permanent injury or death has occurred, the driver can lose the right to drive forever. Germany does not fool around, but does not totally outlaw the consumption of alcohol and driving. This progressive scale of punishment, weighted more heavily towards the habitual drinker and driver, has lowered the percentage of fatal accidents attributable to alcohol to 18 percent on its roadways, only 12 percent on the *Autobahn*. The fact of high speed on its roadways has created a demand to put sober drivers behind the wheel. It is important to note that this does not create an oppressive driving environment. On several trips to Germany over the last ten years, I would quite often drive to restaurants in the numerous small towns that dot the countryside of eastern Bavaria or north-central Austria. During the course of the meal (dining in Europe is a slower-paced event than a quick bite at Arby's or McDonald's in America), I would have a glass of wine before and one with the meal. Following coffee and dessert, driving home was no problem. There was no oppressive fear of being caught or thrown in jail. Despite roadside sobriety checkpoints (I never did have to pass through one), the worst-case scenario would have been a mark on my record if I were a German citizen. There was one thing I would not do if I had had something to drink. I stayed off the *Autobahn* when "impaired" because high-speed driving takes total concentration and clear judgment—the first things to go with alcohol. Such an attitude also moves through the German driving public as it deals with the realities of drinking and driving, along with the safe, prudent use of speed on its highways. Germany has created an effective system of societal and governmental pressure that keeps alcohol use behind the wheel in check, especially in dealing with the habitual drunk driver who causes a higher percentage of alcohol-related fatal crashes.

In May 1998, the blood alcohol level for drunk driving was further reduced to .05. However, this reduction does not carry the same stringent penalties as higher blood alcohol concentrations. The first-time offender pays approximately a $120 fine, receives half the number of points on the driving record as .08 and loses no driving privileges. This new, lower level is an attempt to encompass a broader spectrum of alcohol "impairment." It remains to be seen if this new restriction will improve safety. For a nation like Germany with its societal disdain for drunk driving, this might work. Only time will tell . . .

Cooperation is also a watchword on German roadways. Safety improvements cannot be effective if the government, law enforcement and citizenry are not working together to achieve a common goal: reducing the number of people killed and injured on the highway. To that end, using speed as an incentive, Germany set out to raise its level of seatbelt usage from 50 percent in the mid-1970s to almost 100 percent today. Looking less at the problem of "speeding" on the open road and more to the need to have every driver and passenger strapped into their seats, the government undertook an aggressive program of education, change of laws and law enforcement through the 1980s. In August 1984, a definitive seatbelt law with a fine for those in the front seat went into effect. Compliance rose quickly, jumping from 59 percent in spring of that year to 91 percent by fall. Between 1984 and '85, the number of people killed on German roadways dropped from 10,199 to 8,400—1,799 fewer people. In the following years, belt usage peaked near 96 to 98 percent for drivers and front-seat passengers. Germany has had a greater problem with seatbelt use in the back seat, however. After July 1986, the seatbelt law with a $25 fine was extended to rear-seat passengers as well, but compliance has continued to flutter around 70 percent—an impressive accomplishment nonetheless. A similar program was also implemented for motorcycle helmets. Since it is so easy to spot a rider and passenger without a helmet, compliance jumped to 100 percent within a matter of days.

This dramatic increase in seatbelt usage was the major factor in the decline of fatalities and injuries over the last fifteen years on the German road. For in-town accidents ranging from fender-benders to major pile-ups, what was earlier a crash with fatal or severe injuries became an incident with bumps, bruises or a sore shoulder from the seatbelt—at worst a broken collarbone. Germany had the foresight to require 3-point seatbelts in the back seat of cars as early as 1980. In America, the safety movement was so obsessed with getting airbags into cars, it took another ten years and many back-seat fatalities—even with lap belts on—to get this $2-$3 lifesaver in the rear seat of American cars. Even in a high-speed crash on the *Autobahn* where the traffic is moving in the same direction, many belted drivers and passengers walk away from a crash with only minor injuries. Those who are severely injured are whisked away by an emergency response team of helicopters assigned to the task of getting any accident victims from the crash site to the hospital as quickly as possible. German doctors take justifiable pride in this lifesaving endeavor, but will quickly add that because of the high-speed nature of the road, many crash victims do have very severe injuries—if they survive at all.

Germany treats the average driver on the road as an adult, capable of knowing and comprehending his or her limits behind the wheel. In America, our government and safety movement look down on the average citizen, thinking he or she is too stupid to handle an active responsibility in saving lives. We are, more often than not, convinced that speed should be severely restricted to protect people for their own good—never grasping that this creates an antipathy between the average motorist, the government and everyone else on the road. In Germany, there is trust in government and the average individual. As a result of this different philosophy, increased safety on German roadways has improved dramatically over the last three decades. The number of people killed has declined from 19,193 in 1970 to 6,067 in 1997, a drop of almost 70 percent.

If America had been so enlightened, we'd be talking about 14,000 people dying on our roads today, instead of 40,000. We'd also be saving millions of hours of extra travel time on our interstate and rural highways, further improving economic productivity.

This extensive network of highway safety, and the fact that the *Autobahn* has no speed limit, naturally invites comparisons between road systems in Germany and America. The *Autobahn* represents 2 percent of all roadways in Germany, carrying 29 percent of traffic with 12 percent of fatalities. The interstate is 1.2 percent of U.S. roads, moving 24 percent of vehicles with 12 percent of deaths. Comparing the actual number of people killed on each is impossible because so many more die each year on the interstate than on the *Autobahn.* About 5,000 are killed each year on U.S. freeways while approximately 750 die on the *Autobahn.* The death rate provides a more accurate, but not infallible comparison, between the two. Over the last decade, this rate has flip-flopped back and forth with the interstate being higher in some years and lower in others. The trend of the death rate is more important to look at than any one individual year. In that respect, the *Autobahn* has improved markedly over the last thirty years, falling from 4.35 deaths per 100 million vehicle-miles traveled in 1970 to .74 in 1997. The decline on the interstate has been less dramatic, dropping from 2.69 deaths in 1970 to .88 in 1997. On average, the *Autobahn* carries close to 43,000 vehicles per *kilometer* (6/10ths of a mile) per day, the interstate about 35,000 per *mile.* The length of the two systems must also be considered, with 7,000 miles of *Autobahn* compared to 45,000 miles of interstate. Germany's small size and much denser population puts more traffic pressure on the *Autobahn* system. Its volume has increased dramatically over the last thirty years, up more than 300 percent to the interstate's 200 percent. Some stretches of interstate like I-29 through North and South Dakota are nearly devoid of traffic and accidents. Certain segments of the *Autobahn* around Munich and the interstate around Los Angeles are almost always congested and regularly have fatal crashes on them.

What does this complex battle of statistics suggest? The two freeway systems are all but impossible to compare in any absolute form. Safety becomes a relative term. On the *Autobahn*, the numbers of accidents have been dramatically reduced through all of the precautionary measures taken, even though you are more likely to die or be seriously injured once you have a crash because of the high speed. Safety has been marginal on the interstate, however, because the speed/safety relationship has not been properly addressed. The safety movement in America is only able to respond with its usual, broken-record "speed kills" patter. Safety on the *Autobahn* is much better because Germany deals more openly with the use of speed.

This, of course, leads to the question of speed's involvement in accidents on the German *Autobahn*. How big a role does it play? Taking away the 12 percent of fatal crashes that are caused by drunk drivers (who are traveling too fast at any speed), this still leaves 500-600 fatalities which bear a closer look. A certain percentage can be directly related to bad weather, inattentive driving and a failure to yield, but *all* of these fatal accidents had the contributing factor of speed as part of the reason someone was killed. Germany has worked to reduce the *frequency* of accidents, and to some degree accident severity. Guardrails, the high percentage of seatbelt usage, driver's training and removing drunks from the road all contribute to a lower number of accidents or a reduction in severity. Germany even goes so far as to place windsocks on the *Autobahn* to warn fast drivers against running into crosswinds. Heavy trucks have steel cages underneath their trailers to prevent "under ride" crashes and are banned from the road on weekends.

The problem of speed, however, still remains and is the motivating force for an ever growing number of improvements to the system. It is similar to the philosophy for aviation safety. Speed is a given, and the precautionary measures based on that premise have helped commercial aviation double its speed over the last fifty years while maintaining an outstanding safety record. The same approach is used on the *Autobahn* with impressive results. Passenger cars are routinely driven between 90 and 100 mph, while the death rate is lower than that of the interstate where traffic flows around 70 to 75 mph. Think of how much faster a cross-country trip in America

Death Rate per 100 Million Vehicle-Miles Traveled on the German *Autobahn** and American Interstate Freeway System

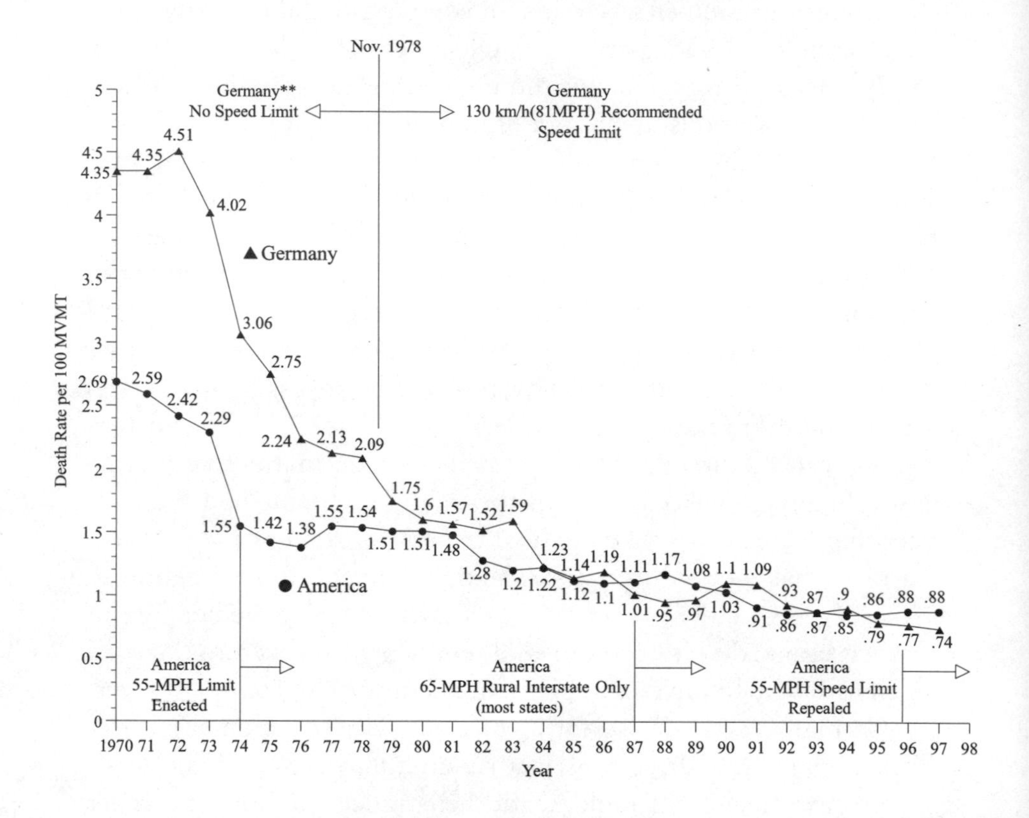

Note: **November 1973 to March 1974, 100km/h(63mph) speed limit on the *Autobahn*.

Sources:
Germany: Bundesminister Für Verkehr
America: Federal Highway Administration

*All figures are for the former West Germany

Travel Speeds on the American Rural Interstate Freeway and German *Autobahn**

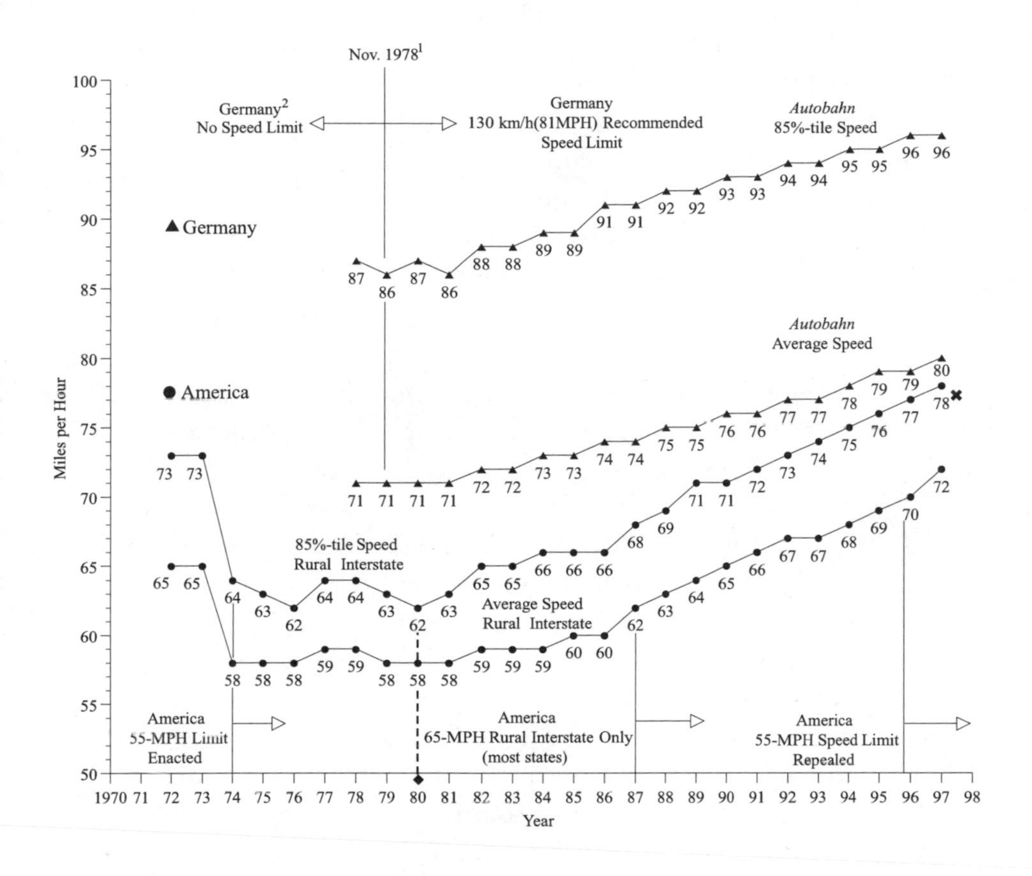

Notes: [1]Speed data before 1978 is unavailable for the *Autobahn.*
[2]November 1973 to March 1974, 100km/h(63mph) speed limit on the *Autobahn.*

◆1980-95 U.S. speeds represent speeds of all vehicles including congested roads, on hills and curves. Free speeds likely 2-4mph higher.

1978-97 German *Autobahn* speeds are for free-moving vehicles.

✖Present-day, free-moving vehicle speeds can vary, plus or minus, 2-4 mph from eastern to western U.S.

Sources:

Germany: Bundesanstalt für Strassenwesen
America: Federal Highway Administration

*All figures are for the former West Germany

would be if these kinds of speeds were both legal and safe? Crossing the state of Nebraska on I-80 could take only four hours instead of the six hours it now takes at a legal 75 mph!

It is important, however, to place Germany's use of speed on its other roads in proper perspective, too. On most rural highways there is a 100 km/h (63mph) speed limit. And given that only a handful of these two-lane highways has a paved shoulder, that limit is quite high. Compliance with this speed limit varies a great deal, ranging from fairly good to quite poor.

Over the last three decades, German vehicle design has improved dramatically in terms of performance safety. It is much easier for these cars to be driven faster than previous models. Given that fact, a growing number of drivers are beginning to exceed the two-lane highway limit by as much as 10 to 15 mph. Germany's "unofficial" response to this problem is to allow speed to go more or less unchecked on rural sections, but to enforce the limit in areas where the highways enter and exit the small towns that dot the countryside. This is as it should be. Most Germans accept and tolerate this kind of enforcement, which is sporadic and unmerciful. You never know when or where it is going to happen, but when it shows up, there is no avoiding it. Radar detectors are illegal in Germany and possessing one results in a stiff fine. The Teutonic version of a speed trap would put the American counterpart to shame. Consequently, speeds are quite free on the open road, but restricted in and around towns. It is a compromise at best. But the extensive network of two-lane highways which crisscross Germany is where the bulk of fatal accidents are occurring today.

For all its apparent love of speed, Germany is still officially uneasy about it. There are already speed restrictions for certain vehicles on the *Autobahn*, starting with the 130 km/h (81 mph) "recommended" limit for passenger cars, light trucks and motorcycles. For single- and double-trailer heavy trucks, there are 80 km/h (50 mph) and 60 km/h (40 mph) limits. These slow-moving vehicles have large red-and-white stickers showing this speed on the back end of the trailer to inform motorists behind them. The fact that many of these trucks and buses have exceeded these limits by significant margins caused them to be mechanically restricted in speed, starting in

1988. For trucks and buses over ten metric tons, this limit is 100 km/h (63 mph), over twelve tons, 90 km/h (56 mph). Unfortunately, in terms of transportation efficiency, this is not working. On stretches of the *Autobahn* with two lanes in each direction, the flow of traffic can be severely disrupted by these slow-moving trucks in the right lane, especially when one pulls out to pass. On the growing number of three-lane-wide *Autobahnen*, this is less of a problem. Trucks can drive in the right lane, and overtake in the center while automobiles pass by, uneventfully, to the left. Still, this mechanical speed limit for trucks should be raised to 120-130 km/h (70-80 mph). This would promote a smoother flow of high-speed traffic that might not be a detriment to safety, either. Most German car manufacturers are also limiting the top speed of their vehicles to 250 km/h (155 mph). Only Porsche and other after-market manufacturers like Ruf, König and AMG are the holdouts, with no restrictions on their cars. This agreement has been quietly reached by most German auto executives, who state that this top speed is *"genügend"* (sufficient), hoping to

Typical German two-lane highway. This Bundesstrasse has a 100 km/h (63 mph) speed limit, and also makes use of boldly painted lines with strategically placed road signs to aid safety at more modest speeds. However, the lack of shoulders on most German two-lane highways along with the possibility of intersection and head-on collisions make for a much higher death rate than on the Autobahn system. American Autobahn Society Archive

stave off governmental regulation or a backlash from the safety community. Given the present state of road and vehicle design, this electronic speed limit seems prudent for the time being.

While my personal experience with the *Autobahn* is completely biased, there should be no doubt in anyone's mind that the German superhighway network is hands-down superior to the present-day American interstate. I say that with a certain amount of sadness because it is frustrating to see such a beautiful freeway like our interstate be wasted on a driving policy that does not maximize its potential in terms of safety and speed. Nonetheless, driving the *Autobahn* for an extended period is an absolute joy, despite problems like congestion and occasional gridlock. The wide-open road is free as the breeze, which is quite noticeable at 100 mph . . .

The *Autobahn,* and European freeways in general, also help to produce some of the most outstanding grand touring automobiles in the world. I felt like a privileged observer of rare and endangered species as, one by one, the exotic cars of the Continent appeared in this high-speed driving environment best suited for their preservation. During my brief time on the road in 1985, '91 and '97, I saw a classic black Lamborghini Countach on the A-6 near Heilbronn, the flash of a BMW M-1 traveling in the opposite direction on the A-8 between Munich and Salzburg, and a certain silver Porsche near Baden-Baden: I had just finished passing a slower car on the A-5 heading north and was moving back into the right lane, when seemingly from out of nowhere, this ageless whale-tailed 930 turbo zoomed past us, charging up a steady rise in the road. We were doing an all-out 110 mph with the Porsche pressing along at perhaps 130 mph. When we crested the hill, all we saw ahead was empty road. The Porsche was simply gone.

Italy is well-represented on the *Autobahn*. Besides Lamborghinis, Ferraris can be frequently seen, but not for long at these speeds. You'll get a better look at one of the many car shows throughout the country during the summer months. I went to one in Bad Füssing near the Austrian border during the summer of 1991. No less than a dozen Testarossas, a half-dozen 308s and even an F-40 showed up.

And this is not just the hoity-toity rich, but automotive enthusiasts who own these supercars and drive them hours on end at high speed—legally—as soon as they leave the car show parking lot. Today, a drive on the *Autobahn* would reveal twelve-cylinder BMWs, Mercedes-Benzes and V-8 Audis. Lucky motorists might also catch a glimpse of McLaren's 240 mph, 630 horsepower F-1, Lamborghini's Diablo or Bugatti's 611 horsepower, 207 mph EB110 Super Sport. As time passes, the names and models will change, but as long as high speed remains a fact in Germany and Europe, these exotics and the spirit of *Gran Turismo*—driving with authority and exuberance—will continue to bring advanced automotive design to the rest of the world.

This begs, of course, the obvious question of how fast is fast on the *Autobahn*? Rudolf Caracciola set a Class B world-speed record of 268.9 mph on the *Autobahn* in a Mercedes Grand Prix racer back in 1938. Bernd Rosemeyer tried to break this record but was killed during the attempt when a strong crosswind caused him to crash on the *Autobahn* in the same year. (This shows the importance of windsocks where crosswinds are present.) There is a memorial to Rosemeyer near the accident site on the Frankfurt/Darmstadt *Autobahn*. For the average driver, speeds of 100 mph are quite common on rural stretches during good weather. Speeds above 130 mph are rare, however. During two weeks of spirited driving at 110 mph in 1985, I saw only a half-dozen cars near or slightly above this speed. In 1991 for a six-month stay, the fastest vehicle I saw was a motorcycle near the 150 mph mark. For a final week of high-speed driving in August 1997, I and a friend were the fastest on the road at 135 mph in a BMW 735—except for one or two cars. Very high-speed driving on the *Autobahn* is just not that common. Specialty aftermarket supercars like Ruf 911 Porsches, AMG Mercedes and König Ferraris with $200,000 price tags and 180+ mph top speeds are almost as scarce in Germany as America. Only rarely do these one-offs and factory specials like McLaren's F-1 and Porsche's 959 find a stretch of *Autobahn* long and empty enough to test out a 200+ mph top speed. At least all of the safety precautions taken in Germany try to make some accommodation for drivers finding out how fast their cars will go. And just because America has lower, restrictive speed limits doesn't mean speeds of this extreme nature don't happen here. In fact, since cars with this kind of potential *are* so rare, this type of very high-speed driving probably happens almost as often in America as Germany. Speeds in the range of 80 to 120 mph need to be addressed more, in terms of safety, than these extremely rare higher speeds.

August, 1997. Over 130 mph . . . in France! Illegal? Technically, yes. The French autoroute system is presently restricted to 130 km/h (81 mph). However, much lighter traffic and, in some cases, superior design to the German Autobahn mean high-speed driving already exists in France, as well as other European nations. Saving more lives in the future will come from dealing openly with that fact, abandoning the use of numerical speed limits in favor of unrestricted-speed freeways with improved safety precautions. Photo by the author

---

If you are planning a trip to Germany to have a go at the *Autobahn*, you need to make a few preparations to ensure that your driving experience is both pleasurable and safe. Start out by obtaining an international driver's license from the AAA. This is cheap insurance, not so much for day-to-day driving, but it helps if you should get into trouble with an accident or have a rental car stolen. Make your car rental arrangements here in the States before you leave for Germany. The rates charged here are dramatically lower. With a bit of persistence you can rent a very fast car, if you learn which firm in Germany the U.S. rental company works with (for example, Sixt is associated with Budget). Even Ferraris and Porsches are available—for a price. Also, check with your insurance agent on coverage. A lot of money is wasted in extra, nonessential insurance that you don't need when you sign the rental agreement. Quite often when you pick up the car, there is a brochure that explains the fundamentals of driving on the *Autobahn*. Take time to read it! Understanding the handful of rules and regulations, along with a dozen or so of the more important international road signs, can spell the difference between flat-out enjoyment and possible trouble. Once behind the wheel, take time to get a feel for the car and its controls. Strap in yourself and everyone else, and don't cloud your judgment and the exhilaration of driving at high speed with *any* alcohol. Believe me, you'll have more fun.

Take your time starting out. It is easy, comfortable and enjoyable to flow with the traffic at 80 to 90 mph (130 to 150 km/h), instead of trying to set a land-speed record right out of the airport exit ramp. Once out on the rural *Autobahn*, open it up as traffic and weather permit, always remembering to pass in the left lane only. This is not just courtesy, but the law. Passing on the right is illegal in Germany and everybody observes this rule of the road, except when heavy traffic makes it all but impossible. If you want to make a run for the maximum speed of your car, wait until the road is completely clear. Don't endanger anyone but yourself in the attempt. This is all common sense for most drivers, but well worth repeating when setting out to drive at very high speed. In April 1991, I had to wait to the end of my first road trip on the *Autobahn* to find a stretch long enough, straight enough and empty enough to do this. Driving southeast on the A-3 *Autobahn* from the Frankfurt airport to the town of Pfarrkirchen in eastern Bavaria, I was near Passau before the opportunity presented itself for a top-speed run. My rented Opel Vectra was not exactly the world-class supercar I would have preferred to be driving, but it was no slouch in the speed department, either. After a brief period of building up speed, the Vectra topped out at just over 125 mph, a speed I was able to maintain for several minutes until I closed in on traffic—*quickly*. Even with intermittent rain and occasional wet pavement, I covered 300 miles in under four hours, averaging right around 80 mph for the trip. On my August 1997 trip, I was never able to find an open stretch of *Autobahn* to peg the speedometer on the rented BMW. The summer traffic was too heavy. I had to remain content with *only* 135 mph...

If you are going to Germany just to drive the *Autobahn* at high speed, plan to take your trip before June or after August. Summer traffic is quite heavy with vacationers from all over Europe using the German freeway to reach their destinations. This makes fast driving all but impossible except for off-peak times. Spring and early fall provide some of the best opportunities to take out a truly fast car and drive it at extremely high speeds (above 130 mph).

You might also want to plan your trip soon. There is an ever growing battle to put some kind of a speed limit on the entire *Autobahn* system—not necessarily for safety, but for pollution concerns. The growing environmental movement around the world, and especially in Germany, is rightfully worried about the problem of automobile emissions. Germany became increasingly alarmed during the 1980s about the number of sick and dying trees in its forests (*Waldsterben*). The environmental Green party was quick to point out that cars, in general, consume twice as much fuel at 120 mph than 60 mph, and thousands of miles of freeway crisscrossing the forests of Germany with no speed limit weren't exactly a benefit to the environment. The more fuel used, the more pollution from vehicle exhausts. The Greens started an aggressive push for an *Autobahn*-wide speed limit of 100 km/h (63 mph). With the battle lines drawn between those interested in the environment and those who want pedal-to-the-metal driving at all costs, the stage was set for a confrontation.

The German government undertook a comprehensive study of the problem in 1985, measuring the effect of vehicle exhaust pollution over a wide variety of conditions from no speed limit to strict enforcement of a 100 km/h limit. I personally drove through many of these test stretches during this period. On the 100 km/h sections, traffic flowed at a reduced speed but nowhere near the limit. Virtually everyone was cheating by 10-15 mph. In areas of intense speed enforcement, drivers reluctantly obeyed the limit, but sped up as soon as the blitz was over (sound familiar?). The final outcome of this experiment was that overall pollution would be reduced by less than 10 percent, and specific pollutants like nitrogen oxide by only 1 percent—not enough to impose a blanket speed limit across the *Autobahn* at that time. Yet both sides became more convinced of the need either for a speed limit or for keeping the *Autobahn* speed limit free. The issue would not die.

The European Economic Community, of which Germany is a part, was also pushing for a Europewide, 120 km/h (75 mph) limit. countries like Holland and Switzerland weren't too thrilled about raising their speed limits to meet the EC's new recommendation, even though many drivers had been pushing for higher speed limits in these countries for years. Germany was consistently showing a lower death rate on its *Autobahn* system than most other nations, so safety wasn't really the paramount issue. Italy's speed limits were a confusing array of progressive speeds, allowing gas-guzzling Ferraris to drive faster than more fuel-efficient Fiats! Germany's and, to a lesser degree, Europe's speed-limit debate ranged from the vociferous to the bizarre.

By the early 1990s, the struggle for or against a freeway speed limit in Germany changed direction. Several legislators who were voted in on the Green's platform of restrictive limits were thrown out by an aggressive lobby of automotive enthusiasts whose slogan, "*Freie Fahrt für Freie Bürger*" (Free speed for free citizens) was beginning to catch on. The fast drivers of Germany were not going to go down without a fight. The environmental movement was also reluctantly having to admit that the reality of an *Autobahn* without a speed limit was moving the nation, albeit slowly, in the direction of cleaner cars and alternative fuels. The freedom to drive at high speed was pushing the issue in a way that slow, restrictive speed limits could not. The average citizen was willing to tolerate higher fuel prices and a move to different forms of transportation as long as free speed remained the law on the *Autobahn*. Mercedes-Benz began running television commercials recommending that Mercedes owners ride a bicycle on short trips, saving the car for longer (and faster) runs. Millions were invested in hydrogen, natural gas and electric alternatives. Every major German auto manufacturer showed off some kind of alternative-fuel vehicle at the Frankfurt auto show. The German car industry claimed as many as 100,000 jobs could be lost if a limit were placed on the *Autobahn*. Instead of griping about the loss of German jobs that a speed limit might, or might not, cause, these manufacturers were quickly moving in the direction of catalytic converters, unleaded fuels and better recycling of the automobile itself. Germany was late in embracing such innovations, but was now making up for lost time. For most Germans, it was difficult to deny that a blanket speed limit would have defused the environmental issue, lulling the average citizen into the complacency of thinking the problem was "solved" while 90 percent of the pollution remained.

Safety was also playing into the speed-limit debate. Improved vehicle design with "safety-cage" passenger compartments to withstand higher-speed crashes was finally beginning to appear in models other than Mercedes-Benz. This was based on tough, new standards for crash protection being phased in by the German government starting in 1998, ushering in a new era where vehicle design would play a more active role in saving lives. Volkswagen's small new Beetle fares much better in front-end crash tests than most American models.

The environmental and safety debates have still not subsided. The election of Chancellor Gerhard Schröder in 1998 and his alliance to the environmental Green party guarantees the issue won't go away anytime soon. It is certain to rage on for years until every vehicle is completely pollution free *and* capable of saving lives in any accident—something that high-speed driving will bring sooner than crawling along at a snail's pace.

Germany is the only nation in the world that has effectively dealt with the problems associated with speed and the automobile. A speed limit on the *Autobahn* would be a serious mistake now that so much progress is being made in the areas of pollution and safety. Germany can take great pride in knowing that it has not stuck its head in the sand, trying to avoid the issue that every other developed nation has refused to deal with openly: The desire of its citizens to drive ever faster, and to do it safely. Germany has held out under growing pressure, both internally and from the world community, recognizing that a speed-limit sign on the side of the road does not guarantee cleaner air, improved safety or obedience to the law.

It's time for America to recognize this same fact, and to begin to deal more openly with its use of speed on the open road.

CHAPTER 4

The Road to an American Autobahn

At present, America is stuck in a quagmire of apathy and hypocritical thinking that prevents us from improving our road transportation system. Such improvements include not only increased safety, but the overall improvement of faster travel combined with fewer deaths and injuries—*true* transportation efficiency. Today's jumble of highway safety programs has far-reaching consequences beyond the obvious applications on the road. They have helped to create a certain transportation stagnation and mistrustful malaise throughout the country—all, of course, in the name of safety. These get-tough policies have had very limited effectiveness in saving lives, while resulting in the increased expense of higher vehicle prices, insurance premiums and longer travel time. The by-product of all this is a them-against-us paranoia that now infects virtually every driver, law enforcement officer and government official from coast-to-coast. Congestion, gridlock and fluctuating economic productivity influence a falling death rate more than a planned attack on the problem. Instead of looking for common ground, the government, insurance industry, automobile manufacturers, law enforcement, truckers, auto enthusiasts and the average driver on the road seek some new scheme to crack down on a safety problem or find a loophole to get around it. All parties involved have shown great tenacity, sticking to their own agendas rather than cooperating with one another.

The common denominator for all involved in this continued lack of cooperation is *money*. Wildcat truckers and trucking firms hope to stave off improved safety regulations to keep their operating costs low. Car lovers arm themselves with ever more expensive and sophisticated radar/laser detectors and jammers to avoid costly tickets and insurance premium increases. Cops struggle to increase funding for more equipment and personnel, striving to maintain power and control over their jurisdictions. The auto industry works to promote the automobile, in all of its manifestations, as the sole mode of transportation for the American people, and fights virtually every safety mandate that might chip into its profit margins. The King of the Hill, however, is the insurance industry, controlling the safety debate with a monetary stranglehold, one-upping the auto manufacturers, beating them at their own game since safer drivers, safer roads, and safer *cars* come at no expense to the providers of insurance. Ironically, insurers fail to realize that using an actuarial table is a lousy way to save lives—everyone is perceived to be at risk. The motoring public ignores all of this, moving down the road ever faster, striving neither to raise the speed limit nor the level of safety on the road. All the while, billions are lost each year through highway fatalities and the inefficient movement of traffic. Government offices and private institutions of almost endless variety spearhead legislation and exercise control over every facet of road and auto safety. The Federal Highway Administration, The National Highway Traffic Safety Administration and all their ancillary branches cost taxpayers billions of dollars annually. Organizations like the National Safety Council, the Center for Auto Safety, Advocates for Highway and Auto Safety, the Insurance Institute for Highway Safety, The National Commission Against Drunk Driving and Mothers Against Drunk Driving, et al, join ranks with this safety conglomerate. Yet, the number of people killed on the road remains basically the same, year after year.

Economic considerations aside, the reason safety has stagnated on the American road is an overriding philosophy that is incompatible with what is happening in the real world: "Slow is

Always Safer" has guided the U.S. highway safety movement since the dawn of the automobile age. Over the last 100 years, motorists have continued to push the needle on the speedometer while safety advocates have struggled to hold it back. This endless tug of war has done far more harm than good. Neither side has backed down in the fight which has become more polarized over the years. With the repeal of the federal 55 mph speed limit, the motoring public can claim a certain victory over the Slow-is-Always-Safer mindset. However, it is a Pyrrhic victory at best. As a nation, we are driving faster than ever before, but there has been no improvement to safety as we've done so—the fault of which rests with a safety community that is out of touch with reality.

There are several fatal flaws with this Slow-is-Always-Safer philosophy. First and foremost is the lack of cooperation. Driving at reduced speed demands a high degree of voluntary support from the motoring public—support which has never been there except for brief periods of crisis, namely the onset of World War II and the first Arab oil embargo. Today, using CB radios, police scanners, detectors and jammers, the driving public continues to increase its speed as the likelihood of getting caught diminishes. Advocates of reduced speed keep stressing that if these devices could be outlawed and enough law enforcement officers placed on the road, drivers could be "forced" to slow down—and live. This is more wishful thinking at best.

It costs about $100,000 per year to pay for training, wages and a squad car to put *one* state trooper on the road. To get the saturation coverage necessary to enforce rigidly today's speed limits on all our highways would require thousands of troopers across the country and *billions* of tax dollars that do not exist. Even if the money were there, it would be extremely difficult, given the prevailing attitude in America, to build the consensus necessary to spend more money to put more police on the road to give drivers more speeding tickets! Further, absolute, strict speed enforcement only guarantees *higher* speed limits. The repeal of 55 proved that. If everyone today got a speeding citation for going 2 mph over the posted limit, the driving public would quickly demand higher speed limits. The reason underposted roads have remained in place for so long is that the odds are nil of getting caught in the first place.

This lack of compliance with lower speed limits plays right into the hands of making money from speed, not improving safety. The Slow-is-Always-Safer movement continues to exploit people's fears on the subject of speed, selectively telling them half-truths. Such "speed kills" rhetoric perpetuates the myth that it is a good idea to hold everyone back. This notion would be laughable if it weren't so counterproductive to safety. Twenty years ago, the general consensus was, "55's a good thing, it keeps people from going 65." Ten years ago, this was modified slightly to 70, then, 75. Today, the chant is, "75's a good thing, it keeps people from going 80!" In reality, the only thing that gets held back is improved safety. This ludicrous situation is kept alive mostly by state governments and the insurance industry. Both strive for speed limits posted below the flow of traffic. More speeding tickets can then be written. States can add the fines to their coffers and pass the ticket information onto insurance companies. In turn, they arbitrarily jack up their rates on millions of customers. This money is not used to improve safety, but is pocketed as pure profit.

Such a state of affairs builds off of the massive, nationwide hypocrisy about safety and speed. How many times have you said, or heard someone else say: "Yeah, well, that's good for all those idiots out there . . ." That, in a nutshell, is the reason we're in the mess we're in. We are all guilty of casting doubt on the other driver while thinking we, personally, are immune from criticism or under no obligation to obey the letter of the law. An environment of mistrust then works through our collective consciousness, insidiously undermining our national resolve to deal with this problem effectively. We look at every other driver on the road as "the enemy" and "the problem." The reality is there are very few "idiots" out on the road. Sure, we all make a boneheaded mistake now and then, but the number of drivers who are completely incapable of raising their level of skill, or have such poor judgment, is very small. It is ridiculous to punish the vast majority of safe drivers with restrictive laws to try and save this tiny minority of "idiots" from themselves. Until we acquire a more enlightened attitude about our fellow driver, we will never improve safety on the road. If you think people are idiots, then force them to act like idiots, some of them will comply. Expect the best, demand responsibility, and you will get it. All our lofty, high ideals of American freedom and individuality don't amount to anything if we apply the lowest common denominator to the average citizen.

The fanatical believers in the Slow-is-Always-Safer philosophy don't want to stop at this point to further their agenda. They still hope some day to implement what they consider their most important safety device. That is the "cost-effective" solution of mandatory speed governors installed in every vehicle on the road. Here, again, is another fantasyland impossibility which lacks any public and political consensus. Even an organization like the Insurance Institute for Highway Safety, which is sympathetic to the idea of a mechanical speed limit, clearly understands that there is no movement in this direction, nor is there likely to be in the future. This doesn't stop them from promoting the idea every now and again, but such a program is hardly cost-effective. Even though speed governors have been around for decades, ranging from a block of wood under the accelerator pedal to high-tech computer chips, retrofitting the 200 million vehicles on the road, from the clunkiest Model T to the sleekest Lamborghini Diablo, would require *billions* of dollars. This measure would do nothing to reduce the majority of death and suffering that happens *under* 55 mph. And what governed speed limit would the driving public tolerate today?

The Slow-is-Always-Safer lobby clings to one, final hope: That they will somehow convince the driving public that "speeding" should have the same social stigma as drunk driving. They resort to heart-rending stories that play upon the public's outrage about a tiny subset of drivers who commit the most outrageous acts. One of the most extreme examples of this appeared in the Insurance Institute for Highway Safety's "Status Report" newsletter. The article recounts the nightmare of a young couple who lost their two children to a maniac driving 80 mph on a *residential* street, broadsiding the family car with his motorcycle. This is the Institute's response to "Why High Speed Matters." Such aberrant behavior on the part of the motorcyclist would be a criminal act anywhere on the planet! It is the kind of story that has absolutely *no* effect in changing people's attitudes about how they drive. The average person on the road would never—*ever*—drive a car or a motorcycle in this fashion on a residential street or condone such driving by anyone else. Fear-mongering based on such abnormal behavior only prevents normal drivers from reflecting on their own outlook towards driving and safety. This tabloidlike exploitation of death and suffering prevents us from looking rationally at an all too serious problem. Likewise, the endless barrage of jingles and catchphrases has no effect in reducing carnage on the road. *Drive Safely, Don't Drink and Drive, Speed Kills, Slow Down and Save, 55 Saves Lives* turn up everywhere in a broken-record litany of worthless warnings. Those who should listen to them don't and those who do don't need to. Average drivers' attitudes will never be modified by trying to associate their "speeding" with the criminal behavior of a small minority of sociopaths or drivers whose judgment is clouded by alcohol abuse. The problems caused by this minority need to be addressed, but will never be successfully resolved until we clearly separate their extreme acts from the actions of the average motorist.

The true bottom line, the one beyond the money, the power and control, is that the American safety establishment has failed—and failed decisively—to reduce dramatically the overall number of people killed on the road. When you look at Germany's steady decline in highway fatalities over the last twenty-five years, in spite of ups and downs in its economy, in spite of the dramatic increases in traffic and speed, that is a stellar accomplishment that makes U.S. highway safety policy look abysmal by comparison. Frantically, American safety advocates look to fringe issues to try and save lives, working to hide the fact that they are philosophically incapable of dealing with the core of our safety problems on the road. "Road Rage" enters the highway safety lexicon for the 1990s. Picked up by the media and turned into an epidemic almost overnight, in reality, it's a problem as old as the chariot—and just about as rare. Putting it into context, anger behind the wheel is as common as the day is long. Violence, from a shaking fist to a punch in the nose, is something most of us will probably experience once or twice in our lives. But homicidal rage behind the wheel happens so seldom, it does not begin to justify the media play it has received. That and "aggressive driving" have been bunched together as reasons for increased accidents and fatalities. This, too, has been given plenty of media exposure as to why things are "out of control" on our highways. (Follow any print or television journalist to the next hot story and you'll know the meaning of "aggressive driving.")

Other larger issues motivate the safety community with equally uninspiring results. They look to stricter drunk-driving laws to save more lives, striving to reduce blood alcohol levels for intoxication nationwide. In this case, if you're not making any new inroads on removing the drunk drivers who are already behind the wheel, by increasing their numbers you can simply round more of them up—whether the problem is reduced or not. In the same vein, the coming use of "red-light cameras" and "photo radar" are looked on as cure-alls for running red lights and catching "speeders." Such high-tech gizmos will, in all probability, simply trade intersection crashes for rear-end collisions, or in the long run succeed in raising speed limits. The real problems of improperly timed lights and using signals to impede traffic instead of move it will go unaddressed. The new euphemism for "Slow Down the Speeder"—"traffic calming"—will have more and more communities sprinkling speed bumps and stop signs where they are not needed, raising the ire of more motorists and the level of traffic violations with little overall improvement to safety. Other more expensive innovations like the creation of turning circles and narrowing of city streets—things that can help cars and pedestrians interact with greater safety—will fall by the wayside. The cost of these long-term solutions is deemed too high, especially when contrasted with the income from fines that can be generated using the cheaper traffic control devices.

Tragically, this is the sad state of affairs which exists on the American road today. The U.S. safety community continues to stumble along, mired in a Slow-is-Always-Safer philosophy that will *never* work. In reality, this philosophy died long ago under the weight of its own infeasibility. The lingering remnants are now costing lives on the road. The motoring public is going in one direction. Safety is going nowhere. This so-called "safety" community refuses to deal with the obvious: If America wants to drive 80 mph, then *true* safety advocates are duty bound to save lives at 80 mph, not lament the fact that people won't slow down, or cynically sit back and say "I told you so" if more people are killed. That's not being a safety advocate. That's being irresponsible.

Ultimately, all these problems speak to the need for a revolution in traffic movement and safety. Taking the first step in that direction requires being open and honest about the risk factors on today's higher-speed roads. While overall fatalities have remained about the same since federal speed-limit control was lifted in 1995, deaths on the interstate system have hit an all-time high. Some of these fatalities can be attributed to shifting traffic patterns as more motorists switch from slower secondary roads to faster freeways, but the increase is there and should be dealt with. The present-day safety establishment can only complain about it. No one is listening and nothing is being done to change it. "Just slow down" is now an incompetent response to a serious problem. We have to move beyond such empty rhetoric—and *quickly*. That requires a whole new approach to the issue.

The faster you travel, the more quickly you reach your destination. This all-too-obvious fact would stand alone if it weren't for so many variables on the road that can become obstacles to driving faster. If you concentrate on the open road, the rural highways and freeways where the issue of moving traffic quickly and safely is critical, you must accept this fact: Speed, in and of itself, does not cause accidents for the majority of drivers on the road. However, as speed in crashes increases, so does the severity of injury and the likelihood of death (this is true whether you're in a car, train or plane—and you don't see restrictive speed limits on the rails or in the air). While an almost endless number of factors determines the outcome of an accident—lucky or unlucky as the case may be—the undeniable reality is, the faster you go, the more deadly it is *if* you are involved in a crash. Unfortunately, this fact gets erroneously translated to "Speed Kills" by many in the safety community. If speed in and of itself killed, we'd all be dead because the earth rotates at 1000 mph! Sudden deceleration is the real killer. This by-product of speed in a crash causes all the death and suffering on our roads. By concentrating on deceleration rather than speed, we can dramatically minimize its effect, saving lives while driving faster. Moving in this direction creates a whole new safety philosophy for today's modern highways: Fast *and* Safe. The only way to improve road safety and transportation efficiency is to make faster driving safer. By merging these two seemingly divergent needs, we have the chance to do some real broken-field running in the battle to save lives.

Achieving an *Autobahn*-style interstate freeway system by shifting our highway safety philosophy away from Slow is Always Safer to Fast and Safe is not an overnight possibility. For that matter, it won't happen in five years. Ten years from today is a reasonable goal, and within the next couple of years, we can certainly make modest increases in speed on our best highways while improving overall safety. If we are to move towards an interstate freeway speed limit and safety record that is in line with that of most European nations, we should be working *right now* for a rural freeway speed limit of 80 mph and a death rate under .80 per 100 million vehicle-miles traveled. States also need to look at the half-million miles of two- and four-lane highways presently posted between 50 and 70 mph, reappraising what the new speed limit should be. By design, these highways show the deadly effects of sudden deceleration more than the interstate, but it is important to look at where the road is located. Are we talking about a narrow, winding two-lane highway in upstate New York with no shoulder and trees planted to the edge, or a wide-open Montana two-lane with little or no traffic and an almost endless line of sight? There is a difference.

Just as Slow is Always Safer comes with a certain credo and philosophical perception of the problem, so does Fast and Safe. Several tenets form the foundation of this philosophy, creating a new network of policies and a shift in national attitude that will revolutionize our approach to safety and speed. If this philosophy is adopted for the American highway, we have a golden opportunity to show ourselves just what we can accomplish when we work together to solve a difficult problem. It requires us to make a number of new assumptions and give up certain familiar arguments in return for a system of precautionary measures that effectively deals with the problem. For the present safety establishment, this means coming to the bargaining table with a willingness to adopt some totally new ways for saving lives on the road. For those of us who wish to drive faster legally, we have to meet the safety movement halfway, realizing that some personal freedom and individual rights must be sacrificed for the common good. Keeping this spirit of compromise in mind, the philosophy of Fast and Safe can be set down and put into practice to improve driving dramatically across America.

1. A clear majority of drivers on the road must be considered reasonable, responsible, *legal* drivers. We will never be able to solve our problems for the long term if we look at ourselves as the enemy. The majority of Americans behind the wheel (in the range of 80 to 90 percent) must be thought of as "safe" drivers. There are close to 170 million licensed drivers in America and 40,000 traffic deaths per year. Most drivers use the system safely. The law must reflect that. This premise alone calls for the reevaluation of almost every speed limit across the country, except for true residential streets.

The 10 million speeding tickets written each year are doing nothing to slow us down or to improve safety. These millions of speeding citations are an indication that the system has broken down (a bad law or insufficient/nonexistent safety programs). Far too many average Americans are receiving tickets, in the vain hope that this overkill will reduce overall travel speeds and reach the few reckless drivers who are also on the road. At the moment, most speeding tickets are given to people who do not deserve them.

The driving experience itself must be made as enjoyable as possible for both fast and slow drivers, as we work to accentuate the positive aspects of the automobile and minimize the negative ones. Give the average driver a reasonable speed limit so that when there is a truly dangerous stretch of road that *does* require a reduced speed, this means something! The American safety establishment continues to perpetuate the half-truth that drivers will always "speed" 10 to 15 mph over any posted limit—something every federal government study over the last forty years does *not* corroborate. Yet, we continue to underpost most roads to preserve an illusion of safety based on misconception, not facts. Witness that the speed limit on a *true* residential street (not a through street) has been posted at 30 mph for decades and there still is 90 percent-plus compliance with this limit because it is reasonable and prudent for the given conditions. Compliance drops dramatically on through streets and other city byways because speed limits are posted too *low*—generally, by 5 to 15 mph. If properly reposted, the issue of speed would quickly limit itself to the open road because travel speeds on these suburban and city streets would be decriminalized, leveling off near the present-day flow speeds of traffic. On our rural highways and, especially, the interstate system, there is some truth to the argument that people will be driving ever faster on a long-term basis. Here, improvements in road and vehicle design, and the absence of pedestrians and roadside obstacles will mean a slow, steady increase in travel speeds over time. Eighty-five percent of the traffic on the *Autobahn* flows in the range of 90 to 95 mph, signaling what we will have to look forward to over the next ten years if we begin to deal with the real-world relationship of speed and safety on our best freeways. The *Autobahn* does, however, dispel the myth that if you put up a 100 mph speed-limit sign, people will automatically drive 110 mph. At the other end of the spectrum, posting a 20 mph limit to encourage people to drive 30 mph instead of 40 mph is nonsense—the average driver simply ignores the sign and drives at a speed that feels comfortable. A prudent, real-world speed limit for the majority means greater trust in ourselves, better compliance with all other laws, and more societal and law enforcement pressure on the minority that chooses to disobey these new speed limits. However, driving faster *legally* will only promote more aggressive driving if there are no corresponding rules, regulations and precautions to permit slow and fast drivers to share the highway with proper courtesy. Our laws must reflect what is expected of fast as well as slow drivers to aid the driving experience.

2. The faster you drive, the greater the accident risk. The speed/ safety relationship is more complex than this statement suggests, but assuming that faster travel is more dangerous allows us to err on the side of caution. Paramount to this premise is recognizing that no one speed should ever be thought of as synonymous with safety. The greatest disservice done by the "55 Saves Lives" campaign was to associate a certain speed with safety. One day on the freeway, 120 mph could be safely driven. Another time, on the same stretch of freeway, 20 mph could be excessive. There is no such thing as a "safe" speed limit or speed. But we don't want to fall into the trap of forcing people to drive faster, either. The best rule of thumb is to encourage people to drive at a speed which feels comfortable.

Related to this premise is the fact that the variation of speed between vehicles is somewhat riskier than a uniform speed between vehicles. Speed variance is a safety problem that must be addressed in a higher-speed driving environment. This still means that in the future faster vehicles will be passing slower ones, but setting out to minimize the accident risk between slow and fast traffic is one more critical step in the right direction. It is vital to recognize in a reasonable, rational way, that speed and speed variance occupy a much smaller place in the overall crash/fatality picture than the safety movement would like to have us believe. However, the risk factor at high speed on the open road should not be underestimated, either.

3. Cars are meant to be driven. They are not and never will be some modified form of public mass transit. There has been endless talk, over the years, of converting our freeways and fleet of vehicles to provide safe, high-speed, computer-controlled transportation. This will *never* happen. It is a pipe dream of transportation engineers and safety advocates who try to defer the real discussion of speed by concealing it in some futuristic scenario that will take faster travel out of the hands of drivers, turning it over to a maze of computers, sensors and grids buried under already existing highways. Even with prototype systems already being tested, this is an impossible undertaking which would cost billions to retrofit both vehicles and highways—money better spent upgrading existing roads and improving alternative transportation in our larger cities. This kind of talk does nothing to improve our attitude about our fellow drivers on the road, but breeds suspicion that "they" are incapable of driving faster. This is one more dead-end, Buck Rogers boondoggle that prevents us, right now, from dealing with the important relationship of safety and speed. Already behind the wheel is a marvelous, God-given computer that has not been used to full advantage—the average American driver.

4. Driving faster and safer requires a certain amount of personal sacrifice. You don't get something for nothing. Working from the premise that we are not going to slow down means taking a new direction to enhance safety. Government and law enforcement can legitimately ask for certain precautions in return for the individual freedom to drive faster on the open road. A law requiring seatbelt use for all drivers and passengers is just a sample of the handful of trade-off laws that will be required for an overall improvement in safety. (Presently, forty-nine states have some kind of seatbelt law, though only eleven practice primary enforcement).

5. Driving under the influence of alcohol must be progressively reduced. We will *always* be both a drinking and driving society. However, the percentage of fatal crashes due to alcohol use can be lowered with a progressive scale of fines, treatment and jail time related to blood alcohol level. This sliding scale will be tipped heavily towards removing habitual, repeat drunk drivers from the road. Under the umbrella of the Fast-and-Safe philosophy, sober, faster drivers—the majority—will be treated as law-abiders. Enforcement efforts can be more effectively targeted against the minority of chemically dependent people who continue to take their problem to the road.

6. A high-profile law enforcement presence is vital to the successful implementation of Fast and Safe. The very thing "speeders" fear the most, the sight of the police, is exactly what is required to oversee the safety measures necessary to maintain a proper flow of high-speed traffic. With a majority of drivers voluntarily obeying the rules of the road, law enforcement time will be freed up to concentrate on *moving* traffic. The outmoded and ineffectual policy of hiding in the bushes—waiting for trouble—will be replaced with police driving with and through the flow of traffic in sufficient numbers, looking for trouble.

7. The interstate freeway system is the only road, by design, that can be adapted for safe, very high-speed driving. Given the layout, access and right-of-way, the interstate already provides safe operation for cars and trucks at modest speeds. Moving passenger car, light truck, and motorcycle speeds into the range of 80 to 100 mph over the next ten years, while allowing heavy trucks and buses to operate near the 80 mph mark, will only be possible on an interstate-style freeway. Because of the inferior design of two- and four-lane highways (the possibility of direct head-on collisions and at-grade intersection crashes), only modest increases of speed in the range of 65 to 80 mph for most roads will be possible. These slightly higher speed limits must include added safety precautions which will help to minimize the accident risk. However, America's extensive network of two-lane highways should also be looked at as alternative routes for slower driving. Care should be taken to post speed limits that reflect the flow of traffic, but an effort should be made to accommodate motorists who travel these roads to enjoy the landscape and who desire to escape the more intense pace of the freeway.

Achieving a high-speed interstate system gives motorists a place to drive fast safely, creating the potential for societal pressure that does not condone high speed in the wrong place. A dangerous, momentary burst of speed on a back road or residential street can be prevented by encouraging drivers to "take that out to the freeway."

For America, the interstate must be thought of as a special roadway, moving traffic at *Autobahn* speeds while improving on the outstanding safety record already established by the German freeway system.

8. Increasing safety on the interstate freeway system has the cascade effect of making all other roads safer. Setting the highest standards of safety and speed on the freeway only improves safety on slower roads. Three-quarters of all fatal accidents happen under 40 mph at intersections and on roads other than the interstate. By using high-speed freeway driving as the basis for *all* vehicle and roadside safety improvements, we improve the safety of slower roads as well. The by-product will be a reduction in deaths and injuries on all roads.

9. The automobile manufacturers and insurance industry must participate in the Fast-and-Safe agenda. Cars and trucks that meet the proper standards of performance and passive safety are crucial to the achievement of faster, safer highways. Due to the perception of danger and the need for technology to solve the problems of a demanding driving environment, a high-speed freeway *moves* safety innovations into the marketplace faster than a slow-speed highway network. The Fast-and-Safe philosophy will help the automobile manufacturers to respond with improved safety features as the law of the land allows their products to fit the performance image of a sleek car streaking down an open stretch of freeway.

As fewer fatal crashes and more lives saved become an ever growing reality, the corresponding reduction in insurance premiums has to be more than just a token benefit of pennies on the dollar. Presently, the "business" of death and suffering on the road is estimated at over $100 billion annually. This loss of productivity, the medical expenses and increased travel time are reflected in the price of our insurance premiums. We don't see this immense cost because it is already ingrained in the system. Better, safer American cars with lower insurance rates stimulate the economy, which is further improved by goods, services and people being moved more safely and faster than ever before. The insurance industry must also stop looking at speeding tickets as a profit-making surcharge tacked on to the premiums of the average driver—which does nothing to improve safety on the road.

10. Driver education must be upgraded and standardized. Improved minimum standards for the training of all new drivers is essential to the overall improvement of long-term safety. A practical program to refresh the memory of drivers who are already on the road, and the implementation of a comprehensive training course for drivers of heavy trucks and high-performance vehicles must work their way into the national safety agenda.

These ten tenets are the foundation of the Fast-and-Safe philosophy. If followed with a steady, unflagging determination, the results would be dramatic. Speeds would increase as fatalities fall, and the average driver would no longer be considered a lawbreaker. Law enforcement would also not be caught in a struggle between the government and the majority of motorists on the road.

Paramount to the implementation of this philosophy is recognizing that no human endeavor is ever perfect. We will never completely solve the problem of death and suffering on the road. Any mode of modern transportation has its lapses in safety. Fast and Safe is not infallible. Human nature and speed are a potentially dangerous combination; but with frankness and honesty, we can minimize that risk, dramatically reducing the number of people killed and injured so driving becomes a much safer form of travel. Using today's traffic statistics as a base line, we need to look at the problem with open eyes so we don't fall into the same old trap of claiming success with a falling death rate and so many lives saved with this or that program. We will never know for sure. Over the short term, dividing the ever increasing number of miles driven by the same approximate number of people killed annually guarantees a falling death rate. A decline in the actual number of people killed and injured from year to year (especially when speeds, miles driven and economic productivity increase) will be the only true indication that we are making progress—as has happened in Germany. Once having achieved this, we must guard against the temptation to stop at some official level and speed limit while, in the real world, people continue to drive ever faster as the grace given by police grows. Allowing real-world speed limits to change over time on our highways and no speed limit for our rural freeways will keep the problem of speed and safety up-front and in the public eye. Some will declare that this, in effect, lets the genie out of the bottle, creating an environment that glorifies speed, the automobile and driving. The genie is already loose. That happened a century ago when Benz, Duryea and Ford built their first prototype cars.

We must stop treating speed as an evil scourge of the open road. Looking at the historical perspective, we see that increased speed has been part of any new system of transportation. From the months it took to cross this country on horseback, to the days by train and, finally, hours by plane, speed was the efficiency that made these improved modes of travel popular, practical and cost-effective. The same must be true for our rural highways and, especially, our freeway system.

Driving faster *and* safer makes sense. We are a big country. Some states are larger than many European nations combined. Here in America, distances are measured in thousands of miles, not hundreds of kilometers. From the standpoint of efficiency, adapting our rural interstate for safe, high-speed operation will be a boon to our national identity, attitude and economy. Unified cooperation in this area means striving forward with a philosophy that provides the potential for great accomplishment. Slow is Always Safer can *never* fit that bill. Imagine if this philosophy had been applied to aviation: No plane would have ever gotten off the ground! Fast and Safe *is* worthy of our pursuit. To take a bad situation and turn it around with hard work, perseverance and cooperation—that is the stuff of historic accomplishment.

But inspiration only gets us started. The devil is in the details.

CHAPTER 5

Performance and Passive Safety: Safe Vehicle Design

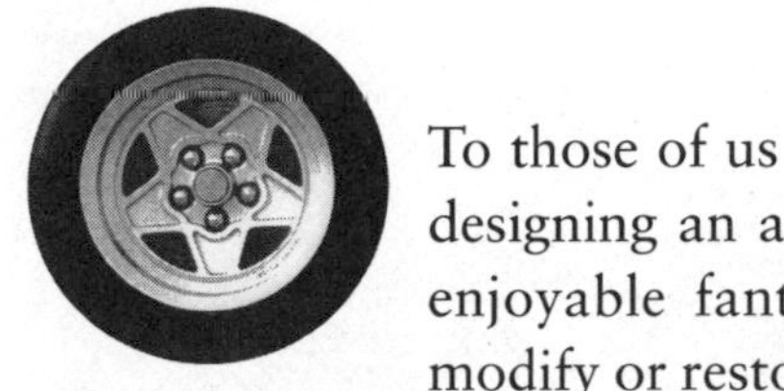

To those of us who truly love cars, the thought of designing an automobile from scratch remains an enjoyable fantasy. Some automotive enthusiasts modify or restore cars already on the road as a way to bring part of that dream to life. But to start with a clean sheet of paper (or these days, a computer screen) and set out to build a car for tomorrow's freeway is a monumental undertaking. Within the confines of today's regulatory environment, the need for safety, fuel economy and lower air pollution seem to preclude anything remotely like driving enjoyment. To their credit, the domestic and foreign auto manufacturers have taken on and met much of this challenge. Over the last ten years, they have produced cars and light trucks that are pleasing to the eye, fun to drive and fairly fuel efficient. To reach this point, however, they have had to endure an onslaught of criticism and aggressive regulation from the auto "safety" movement in and outside of the federal government. This movement has cared little about the enjoyment of the automobile or driving. Instead, it has fostered an adversarial relationship between the auto industry and government that has neither brought the safest vehicles to the road, nor helped the American auto industry succeed in securing a larger market share from the foreign competition.

Auto safety by consumer advocates and government edict.

Performance and passive safety by engineers and business executives who love cars. Bruce Priebe - Twin Cities Section, Mercedes-Benz Club of America

At this point, the focus of concern will not be on this endless battle of regulations or mandates. Rather, attention will be placed on the steps needed to put a car on the freeway that can be safely driven all day long at 80 to 100 mph. The other vehicles on the road will get a design overhaul, too, from motorcycles to 18-wheelers. *All* of tomorrow's motoring needs will be met, taking into account the first and foremost criteria: driving enjoyment. An eye will even be kept on costs and profitability. This is the vehicle you always wanted to drive but couldn't, given the present state of safety on America's highways.

A brief look into the future reveals what you will soon be driving . . .

At first glance, it is one of the most beautiful cars you have ever seen. After all, this is still the main reason why a car sells. The sleek lines and subtle curves should put a smile on your face every time you see them. But it is also vital to feel good about what is underneath the surface styling of your car and to understand why everything is installed and put together the way it is. Driving Fast and Safe means being totally confident when you are behind the wheel. No false sense of security, thinking your car is safe. You *know* it is safe.

Slowly moving around the perimeter, you can see that the ultra-aerodynamic body forms a fluid, uniform shape to cheat the wind. From bumper to fender to roofline, every facet of the car joins to the next, allowing the slipstream of air to rush past with a minimum amount of drag. A look underneath shows that the undercarriage has not been forgotten, either. It is smooth and streamlined, with no muffler pipes or cables left dangling to slow the car down. A tough ABS plastic underliner, similar to those used in pickup truck beds, stretches across the bottom. The entire exhaust system is insulated and isolated down the center of the chassis. The wheel wells, radiator air outlets and tail pipes are the only visible openings to an otherwise seamless undercarriage. In fact, the only parts protruding from the car at all are the two side-view mirrors, but even these are covered with tapered windskirts. This all adds up to an external design that causes virtually no wind noise at high speed and requires minimal horsepower to be driven fast.

The sleek outer body is covered with polymer bumpers, fenders and door panels to reduce weight, as well as the damage from the nicks and dings that happen in parking lots and garages. In minor scrapes of 3 to 7 mph, the "low damage" front and rear end will not require repair, except for a collapsible insert behind the bumpers that can be replaced for under $50.

Popping up the metal hood, you notice that its single-piece, clamshell construction forms both the upper fenders and front end. Along the outer edge, there are several small clips that float in corresponding slots in the fender wells and firewall, giving the front end the maximum controlled deformation in a crash. Beneath the hood is the heart of the car: the motor. With a surprising amount of room in the engine bay, the 90°, V-4, 16-valve power plant occupies very little space, aided by the engine and transaxle being cast as a single unit. This keeps weight and size down without sacrificing horsepower. The 1.6-liter engine puts out 140 hp at 6000 rpms—that's almost 90 horsepower per liter—pushing the car from zero to sixty in eight seconds with a top speed of 130 mph. More importantly, this combination of lightweight, high-output motor and low-aerodynamic drag helps to return 40 mpg at 60 mph, reducing fuel use and air pollution over the life of the vehicle. This means respectable fuel economy even on high-speed trips down the interstate, returning 25 to 30 mpg at 80 to 100 mph.

Everything in the engine compartment is laid out neatly and in its proper place. Even the backyard mechanic can do routine maintenance with ease. With the wraparound hood and compact engine, the drive shafts, disc brakes and radiator are out in the open, easy to reach. The speed-rated tires on 17-inch, light-alloy wheel rims are low profile and squat. The track and wheel base take advantage of the "cab forward" concept pioneered by Chrysler—pushing the placement of the tires out and away from the center of the vehicle. This gives superior tire grip when taking hard corners or sweeping through long curves on the freeway at top speed. The wide tire placement also provides more interior room for the passengers.

Clicking the hood closed and popping open the driver's door reveals an integral part of the overall safety of the car. Almost four inches of padding on the inside of the door forms the interior upholstery. This provides a healthy measure of sound insulation, but also saves the side of your body if a car or truck slams into the door at an intersection. If you could pull the door apart and look inside you would see a laminated sandwich of stamped aluminum and high-density, polyurethane foam. In fact, this aluminum/foam construction surrounds the entire passenger compartment to form a roll-cage. Whether in a front-, side-, rear-end or rollover crash, you are protected from injury with this rigid safety-cell. The front and rear suspension are attached to either end of this roll-cage to provide the tightest frame for hard cornering and to minimize dynamic stress at high speed.

Easing yourself behind the wheel, you find that the seat is fully adjustable, firm, but comfortable, hugging your hips and back with padded side bolsters that wrap around the contours of your body. With the door closed, the driving position is perfect. Visibility is outstanding, with no glare or reflections to obstruct your view. The "A", "B" and "C" pillars (the supports for the roof behind the windshield, front and rear doors) are thin, tapered and padded for a clear line of sight around the car, yet, they are interconnected and act as rollover bars. The headrests for the rear seats are level with the bottom of the back window when not in use, giving you a wide-open view behind. The passenger compartment is divided into four seating areas creating a "safety space" for each person, with room in the rear for a child safety seat. The design gives everyone more room. No one gets short shrifted for leg or elbow room. Even the wheel-well hump where you would place your left foot when driving is gone. A smooth floorboard across the bottom of your feet takes its place. The steering wheel and controls are all within easy reach. You don't have to lean or stretch to press any of the buttons—all of which are flush mounted. There is not one sharp edge or protruding object that is not somehow protected. Even the sun visors retract up into the headliner when not in use.

Sitting back in the seat, you press a button on the right side bolster and the safety-belt straps twist themselves from the side of the seat over your shoulders. No longer a 3-point, two-inch-wide belt anchored to the doorpost and floorboards, this three-inch-wide, 4-point system is attached directly to the back and bottom of the seat. Bringing your hands up flat against your chest, you can easily slip them under and around the straps, clasping the belt latch closed. Another button adjusts the headrest and shoulder straps for the right position for your height. A slight tug on each belt takes the slack out and you're ready for the road. Even though there is a constant tension on each strap, the pressure is smooth and uniform. It's like wearing the vest of a fine suit.

Finally, it's time to start up the engine and head out for an *Autobahn*-style drive. But one thing to attend to first: There is a small lever on the side of the seat that gently locks the seatbelt into place. For the driver, this keeps you in the proper position if you should have to make a sudden move or dive on the brakes. It is not restrictive or uncomfortable, but it does give you that extra edge in an emergency situation. For the passengers—especially those who are a bit uneasy about riding in a fast car—this gives a measure of security and comfort that is not false or mistaken. The integrated seatbelt in combination with the roll-cage design gives you more than a fighting chance to walk away from almost any accident on the freeway.

As you engage first gear, the car moves smartly out onto the road. The extra-wide track is nimble and easy to push through corners or lane changes. The 5-speed manual transmission has crisp, clean shifts, thanks to a short-throw gear linkage that uses brass bushings and spherical rod ends to connect it to the gear box. The ratios are well matched to the torque, horsepower and high revs of the engine—just as they are on the automatic version. A dropped gear and accelerator pedal to the floor push the engine smoothly and quickly to the 8000 rpm redline. The on-ramp is easily negotiated, as is merging onto the freeway. The miles slip by in a relaxed and easy manner, no more fear of the police or annoying buzz of the radar detector. Resting comfortably in high gear, it's time to cover up the instrument cluster and find out which speed is just right for cruising on a wide-open stretch of superhighway. After several minutes, the speedometer indicates a shade under 100 mph. It's not taxing in the least. The steering and ride are precise, firm and taut. Wind noise is almost nonexistent. The tiniest details spell safe driving: From the perfect position of the arm rests and steering wheel, to nighttime driving with side and rearview mirrors that don't reflect glare in your eyes, safety at speed is paramount. The headlights now have three settings: the normal low and high beam, and an extra bright, narrow beam for high-speed freeway driving—actuated by pushing the turn-signal stalk an extra "click" forward. Day or night, the double wishbone, four-wheel, independent suspension and disc brakes with antilock control can quickly haul you down from the highest speeds if someone should suddenly pull out in front of you. And for the driver and passengers who are not "locked in" with their seatbelts, the harness system automatically senses the change in speed, firmly restraining everyone before an inch of slack pays out—taking advantage of the latest antispool seatbelt technology. The bottom line is total driver and passenger enjoyment, complete security and comfort under all driving conditions.

As you slowly accelerate beyond the triple-digit mark, the sides of the road rush past and your concentration becomes more focused. You can still sweep the car through the long curves of the freeway with no trouble, planting it firmly in the center of the lane. As the road straightens out, the speedometer needle edges ever closer to 130 mph. As you finally reach top speed, the wind noise and whine of the motor hit a pronounced, solid moan. This is a handful. Exhilarating, but intense. Backing off, the car settles down to a more comfortable 90 mph. Not a bad test drive for the standard automobile from Detroit—all for a shade over $16,000. And this is not an econo-box or a sports car, either. The seats are comfortable, the interior spacious and uncluttered. It's a driver-friendly environment, whether you are puttering your way to the corner drugstore or driving cross-country on a high-speed trip. If this sounds a bit like a commercial, that's what it's intended to be. Performance and passive safety *can* sell, if the car is packaged properly and the carrot of speed is dangled safely in front of the customer.

This road test is, perhaps, five to seven years away, but many of the innovations discussed are already here today. A closer look at the features that have not made their way into the average American automobile are in order, to understand better what it will take to get them on the road as quickly as possible.

This car of tomorrow has to accomplish so many things if it is to fit properly into our future road transportation system. Safety, pollution control and affordability must be connected to every facet of its design, allowing a specific innovation to solve as many interrelated problems as possible. When a design change can enhance the overall performance of a vehicle at many different levels, such improvements are then more cost-effective and have a greater likelihood of reaching the marketplace sooner.

The improved aerodynamics for our car of tomorrow accomplish several things at the same time. The smooth, fluid shape, which dramatically reduces the drag from wind resistance, also creates a friendlier exterior if a pedestrian is ever struck by this vehicle. This is not to say that a person could not be killed or severely injured when hit, but this advanced design with no protrusions jutting outward helps to minimize injuries. On the freeway at speeds above 80 mph, such aerodynamic refinements reduce fatiguing wind noise, provide greater vehicle stability and require less horsepower to be driven faster. This translates into safer, more enjoyable driving that uses less fuel and produces lower air pollution. As speed increases, the drag created by the wind being pushed past the body becomes the single biggest force to be reckoned with. The drag created by the rolling resistance from tires and moving parts is minimal by comparison. Reducing this coefficient of drag means you have opened the door to higher speeds with fewer penalties. One innovation—improved aerodynamics—effectively deals with several requirements of future vehicle design.

Closely related to the problem of wind resistance is the engine powering tomorrow's car. Higher speeds, lower fuel use and pollution control require the continued downsizing of the *standard* passenger-car engine. This still means a Lamborghini will have twelve cylinders or, for that matter, a Corvette a V-8—these cars represent a small number of total-performance machines that help to pioneer better technology for all vehicles. For the typical family sedan, however, a four-cylinder engine of 2 liters or less is all that is needed for future high-speed driving. With proper dynamic balancing and tuning, such a motor can still provide plenty of smooth passing power and acceleration. Turbocharging, multiple valves per cylinder and variable timing already boost power without affecting gas mileage. Combining the engine, transmission and differential into one compact unit would further decrease size and weight. Expanding on the use of aluminum, plastics and composites will also reduce the weight of the engine. All of these present-day innovations help to tweak more horsepower and fuel economy out of ever smaller motors. Over the next decade, such improvements will find their way into a larger percentage of automobiles, helping to reduce our overall consumption of oil and further minimize damage to the environment.

Clean exhaust emissions from tomorrow's internal combustion engine will remain a challenge. However, already on the market are

LEVs (Low Emissions Vehicles) and ULEVs (Ultra-Low Emissions Vehicles). Quickly growing in number, these cleaner-burning motors will help to hold down the levels of carbon monoxide, unburned hydrocarbons and nitrogen oxide. In the coming decade, however, we will have to take a hard look at the best short-term, cost-effective improvements for cleaner air across the country.

The Environmental Protection Agency (EPA) is striving to improve air quality by mandating cleaner-burning gasoline and ever lower emissions for new vehicles and by implementing "enhanced" inspection of motor-vehicle emissions around the country. The fact that new-vehicle emissions are already quite low—some 97 percent cleaner in terms of hydrocarbons than the ones built thirty years ago—means a 1 or 2 percent reduction in pollution will be minuscule at best. And most vehicles on the road will already pass the new and expensive dynamometer test the EPA has in store. However, a short-term, cost-effective improvement to air quality *does* exist. It means shifting our emphasis away from new vehicles and blanket inspection programs to certain cars and trucks already on the road. A study done by the state of California in 1991 concluded that 10 percent of its vehicles produced 50 percent of the pollution. This was not the antique or collector car in good tune and repair, either. It was a broad spectrum of vehicles, ranging from the old clunker with leaky valves to the late-model sedan with emissions controls out of sync. By going after this one-vehicle-in-ten that is a "gross polluter," we would more cost-effectively reduce auto exhaust emissions. This improvement does come with difficulties, however. Most vehicles can beat an emissions testing program by using various gasoline additives in a hot engine. A maximum dollar value for repairs is usually set by states that require such a test. After the limit is reached, the vehicle is free to pollute as much as before.

The money presently being used to try and reduce overall emissions would be better spent repairing these high-polluting vehicles or removing them from the road. Already, there are roadside sensors to measure vehicle emissions. Such equipment can be placed on heavily traveled roads. The suspected gross-polluting vehicle has its license plate automatically photographed and a letter is sent requiring an emissions test. Most vehicles can be cleaned up for under

$500—which could be paid for from the savings from today's inefficient testing programs. Those who refused to have their vehicles tested and cleaned up would not receive their license tabs until repairs were completed. Law enforcement could also play a role in spot-checking for vehicles trailing long plumes of blue smoke down the highway. Such monitoring makes more fiscal sense than further regulating new cars or testing the majority of vehicles on the road. Other "solutions" like alternative fuels and electric cars are still, unfortunately, a long and expensive way off. Attacking 50 percent of our pollution problem in this manner would bring a more immediate return on our investment in clean air.

A properly designed running gear is the foundation for safe driving at any speed. Improved steering, brakes, tires and suspension can aid driving enjoyment and help motorists *avoid* accidents. Over the last decade, this facet of performance safety has been refined to a high degree by all automotive manufacturers. The average new car handles lane changes, cornering and moderate high-speed driving (in the range of 80 mph) with positive feedback through the steering wheel and brakes. Present-day performance sedans and sports cars have attained such a level of refinement that many can be driven all day long at speeds above 100 mph and not be the least taxing or fatiguing to the driver. Such running-gear improvements have probably done more to bring about the slow, steady increase in U.S. highway speeds and speed limits than any other factor. The average family sedan "feels" comfortable and safe at 75 mph and most motorists are simply taking advantage of this reality.

In the movement toward driving faster legally, the natural trend will be the continued refinement of road handling to promote comfort and stability at ever higher speeds. Improvements to the basic package of four-wheel independent suspension, disc brakes and rack-and-pinion steering will move tomorrow's average car in the right direction. Driving enjoyment will be balanced between reduced effort to control a vehicle at higher speed and enhanced evasive capability to brake and steer away from a potential accident. Such improvements in running-gear design will create a more direct connection between the road and driver. The individual behind the wheel can then better respond to prevailing road conditions, being more confident that the input from the steering wheel and brakes translates to increased control of the vehicle itself.

Several running-gear innovations have the potential to improve this link between the road and driver. The first is the tread that the tires put down on the pavement itself. Road-holding is only as good as the traction available from the amount of rubber in contact with the type of road surface. Over the past quarter century, radial-tire technology has been dramatically refined in terms of traction and durability. Tire failure due to overheating and blowouts has become increasingly rare. But today, the standard tire on the American road lacks one small requirement: a speed rating. Long since used on the *Autobahn* and other European high-speed freeways, speed-rated tires provide better control for reducing heat from rolling friction and retaining a uniform shape as speed increases. These tires are rated by letter to a certain mile per hour, indicating the maximum safe operating speed. Overtaxing tires by driving them faster than their rating is a safety concern that needs to be addressed. Just as proper width, diameter and "grip" of a tire are important to consider, so is the ability to handle the top speed of a vehicle. Though most drivers over the coming years will not operate their cars, trucks or motorcycles at top speed, a growing percentage will. Speed-rated tires sold as Original Equipment Manufacture and as replacements will give a small, but added margin of safety by reducing tire failure due to high speed. And the specter of a federal regulation and a lengthy battle to mandate them does not have to be a part of the process, either. More and more vehicles are already being equipped with speed-rated tires for just a few extra dollars per wheel. Sanctioning faster driving will ensure that they become a standard part of the purchase price, without the bureaucratic nightmare to require them.

Another innovation is a set of interrelated mechanical and electronic devices to manage better the relationship between tire traction and the road surface. Antilock brakes, traction and antiskid control have become big selling options for all automotive manufacturers, enticing customers to spend hundreds, even thousands of

TIRE SPEED RATINGS

Rating	Certified Top Speed
S	112 MPH
T	118 MPH
U	124 MPH
H	130 MPH
V	149 MPH
W	168 MPH
Y	186 MPH
ZR	Beyond 149 MPH

dollars to give them an added edge in controlling their vehicles in an emergency situation. The antilock brake system (or ABS for short) uses a sensor in each wheel to monitor its rotation while a computer modulates hydraulic pressure in the master cylinder, brake lines and disc-brake calipers. In a panic stop, the wheels are held right at the threshold of locking up by pulsing the brakes several times per second. The driver can maintain better steering control because a wheel that is not "locked up" will skid less, providing more directional stability. With traction control, the process is reversed. As soon as a tire begins to spin, the brakes momentarily engage as engine rpms automatically drop to bring it under control. Restricting wheel spin puts a slower tire in better tread contact with the road surface, improving its "grip." For antiskid control, sensors measure steering input and yaw motion, reducing the possibility of a vehicle spinning out by activating the left or right rear brake, depending on the direction of the skid.

In the relationship between faster driving and safety, antilock brakes play a more important role than does traction or antiskid control. However, this computer-enhanced brake system does not presently provide improved stopping power under all driving conditions. Its advantages and limitations must be better understood so it can be properly integrated into tomorrow's vehicle design.

Through manufacturer's claims and advertising hyperbole, the effectiveness of antilock brakes has been somewhat exaggerated, and many drivers have incorrectly assumed what this system can and cannot do. It is vital to remember that you will never have more traction than what already exists between the tire tread and road surface. Antilock brakes only permit the driver to better *manage* the relationship between brake effectiveness, steering control and tire traction. In simple terms, you can have a lot of braking or steering but not both. Computer-controlled brakes only activate when one or all of the wheels approaches lockup. In most systems presently on the market, the brake pedal "pulses" heavily against the driver's foot in the antilock mode. Many motorists have found this disconcerting and have actually backed off the brakes, thinking something was wrong. Others have continued to manually pump the pedal, further reducing antilock brake effectiveness. In addition, when the driver turns the steering wheel to avoid some obstacle while the antilock system is engaged, the brakes automatically release to permit the wheels to turn. The more steering input, the less braking ability. This effect has been so pronounced in some accidents that drivers have claimed their brakes failed entirely, when subsequent investigation revealed they had not. The steering input had been so great as to cancel all or most of the brakes' ability to slow down or stop the vehicle.

Here is where education and advertising responsibility can influence our ability to take advantage of this technology. In an emergency evasive maneuver which requires both maximum stopping power and steering control, as much straight-line braking as possible should be done first. The average motorist's inclination is to slam on the brakes, which activates the antilock system. Then, as the vehicle slows down, the driver steers away from the potential accident at the last moment. Since panic stops rely on split-second reactions, this maneuver may be difficult to teach to some drivers, but at least the assumption will not be made that the antilock system can permit stopping and steering with equal ability.

On all types of pavement, wet, dry, ice or snow covered, vehicles equipped with antilock brakes can be kept under better steering control by the driver. With conventional brakes, the tendency would be for the brake system to lock up, causing a skid or spinout. The vehicle with antilock control would retain better steering control than the one without. However, these computer-controlled brake systems do have their disadvantages. On loose gravel, lightly packed snow and certain types of extremely smooth glare ice, conventional brake systems can actually stop shorter than antilock ones. Under such road conditions, a tire that locks up will "dig in" to reach the parent road surface underneath. The tires on an antilock system will continue to ride on the upper, looser or smoother surface, taking longer to stop. Steering control would still be more effectively retained with the antilock-equipped vehicle, but if maximum brake effectiveness is what you want—the ability to stop the vehicle in the shortest distance—antilock control does not always provide the best result. That is why there should be an antilock cutout switch on the steering wheel or dashboard to permit the driver to choose between a computer-controlled *or* conventional brake system. Under normal driving, the antilock system would always be engaged. On a road surface with loose gravel or snow, the driver could disengage the system to take advantage of conventional brake lockup. Once the vehicle has stopped, the antilock system could automatically reset itself, ready to provide the best blend of braking and steering control until overridden by the driver. This more comprehensive brake-management package would take better advantage of the skill of the driver and the vehicle's ability to provide the best overall stability and control.

Even though antilock brakes don't solve all of tomorrow's stopping problems, they do have one clear-cut advantage over conventional brakes on vehicles driven at high speed. Up to about 30 mph, conventional and antilock brakes take about the same distance to come to a straight-line stop on dry pavement. As speed increases beyond 50 mph, antilock brakes show a marked, improved ability to slow down a vehicle in a shorter time period and distance than conventional brakes. This is due to the computer system's ability to draw the brakes almost instantaneously right to the edge of lockup, more effectively taking advantage of the tire/road traction available. With conventional brakes, the tires begin to skid over the pavement the moment they stop turning, dragging across the road surface on one small area of the tire. The antilock system permits the tread around each wheel to be in more continuous contact with the pavement, slowing and stopping the vehicle more quickly.

For tomorrow's faster-paced freeway, such an advantage needs to be exploited more fully. The reality of slower vehicles pulling out to pass as faster ones zoom up from behind will inevitably increase with speed. This requires more effective slowing-down and stopping power. The standard car of tomorrow should be able to a stop in a range of 125 feet from 60 mph, approximately 240 feet from 80 mph. This is for straight-line, dry-pavement braking with no steering input. Many cars today already come close to these figures. It is important to note that the actual stopping distance is less critical than the antilock's ability to reduce vehicle speed quickly from, say, 80 to 60 mph to minimize or avoid an accident. For this reason, drivers with a taste for speed should seriously consider having antilock brakes on their vehicles—and learning how to use them.

If the correct combination of brakes, suspension and steering forms the foundation of road-going safety, then they must be attached to the vehicle's chassis, which handles the dynamic stresses of driving, provides proper interior ergonomics and protects passengers in a crash. Here is where performance and passive safety merge into one. The by-product of this union is roll-cage or safety-cell construction around the passenger compartment and fuel tank. This is the tough, resilient framework of the doors and vehicle body that will not crush inward on the occupants. It is this protective "cage" that saves your life, even in a severe crash. Many manufacturers already advertise that their vehicles have this type of design, but such claims can be misleading. Many of these "safety-cells" are anemic at best. Most new cars and trucks on the road today have substandard protection for rollover crashes, allowing the roof to collapse in on your head and neck. In certain kinds of off-center, front-end collisions, the steel around the feet of the driver and front-seat passenger bends inward, causing broken ankles and legs. In fact, only in a Mercedes-Benz or Volvo is the passenger compartment rigid enough to be called a true roll- or safety-cage, protecting everyone in all types of collisions. In our efforts to bring about tomorrow's Fast-and-Safe freeway, such safety-cell construction must be required of all automotive manufacturers.

Presently, most automobiles are constructed by using a patchwork of stamped-steel and spot-welded panels that forms a "unibody" subframe. Attached to this structure are the doors, body panels and running gear. Such a design provides adequate strength for the forces involved in driving the car, even at high speed, but does a rather poor job of protecting the driver and passengers in a crash. While the crumbling of the vehicle's front and rear end is desirable in a collision—absorbing part of the impact—this crushing inward of the vehicle quite often extends right on through to the passenger compartment, injuring or killing those inside. If, on the other hand, everyone were properly restrained from striking the interior, and the passenger compartment did not collapse in on the occupants or fuel tank, the result would be a crash with no injuries or death—even at speeds above 100 mph on the freeway.

Moving in this direction means, first, filling in the gaps that presently exist on today's "safety-cages." The second is steering future vehicle design towards new materials and construction techniques that improve roll-cage strength while holding weight and costs down.

In the short term, a small amount of steel needs to be added to the front foot well and roof of most vehicles. This would tie together the side-impact protection that already exist in the doors, more completely surrounding the passenger compartment with a basic roll-cage. For the 50 percent of accidents that are front-end collisions, this would further reduce the rearward movement of the vehicle's frame into the legs and feet of the driver and front-seat passenger. This is especially important in "offset" crashes where two vehicles collide left-front headlight to left-front headlight when traveling in opposite directions on a two-lane road. Such accidents place great strain on the left side of the vehicle, which easily crushes inward on the driver unless the area behind the engine firewall is properly reinforced. Bracing and tying together the roof pillars with extra steel further restricts them from collapsing, helping to reduce the 20 percent of fatalities caused by rollover accidents. In side impacts, the extra rigidity of the roof and foot well defuses crash energy around the occupants, curving the vehicle's body like a banana. Three to four inches of interior padding on each door further improves the survival rate in such collisions. This addition of metal and spot-welds would add about $100 to the price of an automobile—relatively cheap insurance against the possibility of such crashes. However, auto manufacturers are not fond of revising models already moving down the assembly line. While such patchwork improvements can be phased in, given enough lead-time, more fundamental design changes are needed in vehicles now on the drawing board.

The present unibody crushes inward on the passengers because the stamped-steel and spot-welded subframe forms a thin, hollow structure around the passenger compartment. Simply using a heavier gauge steel and more spot-welds is not the solution to the problem, either. The vehicle would become prohibitively heavy, requiring a larger motor and brakes to handle the extra weight. Taken to the absurd extreme, you would soon be driving down the road in an armor-plated tank. Thankfully, one answer to this problem is a structural polyurethane foam that fills in the hollow spaces of the unibody subframe. This foam can be packed so tightly together that it is almost indestructible, yet it remains extremely lightweight. It still, however, must be used in concert with the stamped steel of the unibody to provide the proper structural rigidity for driving. The foam by itself cannot handle the flexing and twisting the frame undergoes on the road. It is this blend of steel and foam that bonds performance and passive safety together. The unibody would now be extremely rigid, giving the best road-holding response for high-speed driving, hard cornering and heavy braking. In a crash, the front and rear end would still retain their ability to crumple and absorb the force of impact, but this controlled deformation stops short of the passenger compartment. The foam/steel roll-cage better restricts the

All cars are not created equal. A Mercedes-Benz CLK coupe in a front-end, offset crash test. Notice no car body intrusion into the passenger compartment. This kind of structural integrity, combined with an excellent seatbelt restraint system means that airbags can act as true supplementary protection, dramatically improving your chances of walking away from the most severe collisions. Courtesy of Mercedes-Benz AG

inward movement of the car body toward the occupants. The Infiniti Division of Nissan Motor Corporation started moving in this direction in the early 1990s, using structural polyuerythane to enhance unibody strength around the windshield and front door frame of its J 30 sedan. The German car manufacturer, Audi, has a fiberglass and foam insert in the windshield frame of its convertible Cabriolet to improve rollover protection. Over the next decade, such use of foam and other composite materials like Kevlar and carbon fiber will become common in more vehicles as a lightweight alternative to improve frame strength and crash protection.

Roll-cage construction would make the growing percentage of minivans, light trucks and sport utility vehicles some of the safest vehicles on the road. Their more squared-off dimensions and generally larger size would give an extra level of crash protection to those inside. Having lower bumper heights and more deformable front-end subframes on these vehicles would help lower the number of deaths and injuries in collisions with smaller vehicles, too.

Making this changeover more cost-effective for auto manufacturers will help bring roll-cage construction to the marketplace sooner. Such savings could come by switching the assembly line from spot to seam welding of the unibody. Instead of a collection of steel panels held together by thousands of electronically fused "spots," each panel would be welded to the next along its entire length. From bumper to bumper, the vehicle frame in effect becomes

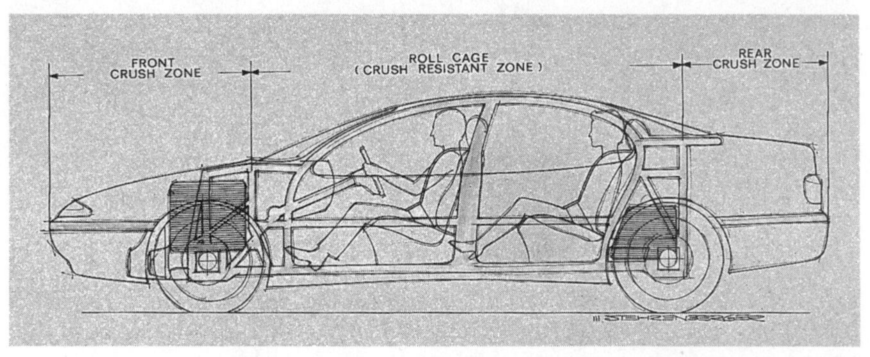

Front and rear "crumple zones" are already standard for virtually every modern vehicle on the road. What is still lacking for most is a roll-cage which protects all occupants in every kind of crash: front, side, rear end and rollover. Illustration by Mark Stehrenberger

one solid piece. Welding all of the seams in this fashion improves the strength of the roll-cage even without the foam. It also makes for a quieter ride, eliminating the squeaks and rattles that develop over time from the constant flexing of the individual steel panels and spot-welds.

Phasing out hand-held spot-welders while converting robotic ones to seam welding is another way of saving money in the long run. A change like this could mean a substantial reduction in the amount of steel used by a car company. A spot-welder requires a "lip" on each panel about a quarter inch wide to join it to another. For seam welding, this lip can be reduced almost to nothing, just wide enough for the electrode to fuse the panels together without the aid of a welding rod—another cost savings. Even taking into account the extra steel needed to reinforce the roll-cage, a quarter inch less on each part of the unibody is an enormous savings to a car company when considering the hundreds of thousands of vehicles built annually.

Integrating the high-density foam into tomorrow's vehicle design is more difficult. Its initial purchase, molding and assembly will come with a higher price tag, but some possibilities to streamline the assembly process are worth exploring. Premolding the foam could create a rigid, ready-made framework for the steel panels to be snapped onto, holding them in place for welding without the elaborate jigs presently used. Or, this process might be reversed, welding the panels first, then injecting the foam into the frame. Weight and cost could be reduced by finding the right ratio of steel to foam, determining the proper gauge and geometric shape of the steel to the correct thickness of the foam. Such considerations open the door to improvements that don't necessarily cost the manufacturer or consumer more money. Clearly, options do exist to modify vehicle design in this way without adversely affecting purchase price. While the initial investment would be sizable, such changes should bring a healthy return to enterprising car companies that have the vision to explore these innovations. Further incentive would come from the real and perceived dangers of faster driving, creating the "need" for vehicles to have rigid roll-cages. Buyers might even be willing to pay more, if necessary, to have them.

If performance and passive safety become one in roll-cage design, the focal point of this merger is the driver's seat. No other part of the car has such all-encompassing importance for safe motoring. The standard seat in today's cars and light trucks provides only adequate support for proper driving. In performance terms, it will hold you reasonably well for modest cornering or a panic stop. During an emergency maneuver to avoid an accident, most drivers will pitch back and forth in the seat but still retain control of the steering wheel and brakes. This, however, is true only for bucket seats with a certain amount of bolstering on the side. The old-style bench seat provides no such support. You would simply slide away from the wheel, with a much higher probability of losing control.

In terms of crash protection, the average seat rates even more poorly. For rear-end collisions, the headrest is often too low or too far back in a fixed position to brace the head and neck properly, and still allows "whiplash" to occur. The back of many front seats can break lose in a rear-end crash, severely injuring the person in the seat or the individual behind. In side impacts, the head and upper body pitch toward the point of collision, striking the roof, door frame or window glass, not to mention the other occupants in the vehicle. During a head-on crash, the driver and passengers can slide right out of their seats—even with seatbelts on—because the bottom cushion in many vehicles is too flat, permitting them to "submarine" right under the lap belt.

Such shortcomings extend to many safety-belt systems in today's vehicles as well. Some older style belts have a "lampshade" or ratchet mechanism that can allow several inches of slack between you and the belt. In a crash, you could be halfway through the windshield before the safety belt begins to restrain you. Thankfully, this type of system is being phased out, but millions are still on the road, providing only marginal lifesaving ability. On most modern seatbelts the shoulder strap is wound around a reel which locks into place by the inertia created during sudden deceleration and/or change in vehicle direction. In a collision, this nylon belt webbing tightens around itself on the reel, in some cases allowing several inches of slack to pay out after the mechanism locks. This "spooling" of the belt can dramatically reduce its effectiveness as crash force increases. Automatic, motorized belts are the worst of all. Mandated to force people to wear at least a shoulder strap, it is all that many drivers and front-seat passengers ever use, if they don't disconnect it. Using only the shoulder strap gives you meager restraint for heavy braking, and none for an evasive maneuver. In a front-end collision, you'll hang yourself by the throat against the strap if you are not wearing the lap belt. A few systems are even attached directly to the door. If it pops open in a crash, you'll fall right out of the car. To make matters worse, many seatbelts don't fit tall or short people. The two-inch-wide strap pulls and chafes against the upper body, and the lower belt rides up on the hips. The final outrage is that some center push-button seatbelt latches actually pop open in a crash! This leaves no possibility to save lives or reduce injuries even for people who have taken the time to buckle up. For the 60 percent of Americans who presently use them, the lifesaving potential of seatbelts ranges from quite good to nonexistent.

The majority of 3-point seatbelts installed in cars and light trucks over the last decade *do* provide adequate protection. However, for the seats and safety belts of tomorrow, a more uniform design needs to be developed to aid motorists in faster driving, avoiding accidents and surviving crashes. Fortunately, several manufacturers are working in the right direction.

Mercedes-Benz and BMW introduced seats in the early 1990s which incorporate the safety harness into the seat itself. Such integrated design optimizes both comfort and safety. Support for your back, legs and hips comes from firm padding and wide side bolsters, which help to minimize fatigue on long road trips. Cradled "in" the seat rather than perched on top of it, you can better handle the physical stress of sweeping the car through high-speed curves on the freeway. During the first instant of an emergency situation, your body is held more effectively in place to retain better control of the steering wheel and brakes. In all types of collisions, the wrap-around design of the seat and belt minimize body movement, reducing the possibility of striking the interior. The upswept seat bottom prevents you from sliding under the lap belt. Crash energy is more effectively absorbed through the reinforced seat back, its strong frame acting as a mini "safety-cage" around your body.

The Mercedes-Benz seat has a cylinder of sodium azide (the same compound which inflates airbags) attached to the seatbelt reel. In a crash, this device is ignited, creating the instantaneous pressure to take up several inches of slack in the belt, restraining you more effectively. The seatbelt system in BMW's 850CSi coupe has a "webbing-grabber" mechanism at the shoulder strap outlet, also helping to reduce the "spooling" effect. Unfortunately, these early integrated seatbelt designs had only one drawback. The Mercedes-Benz seat with its rigid, but lightweight magnesium frame, pyrotechnic seatbelt pretensioner system and electric motors to adjust it costs more than *$8,000*. Fortunately, other manufacturers from Buick to Chrysler are now offering integrated seatbelts for much less. This opens the door to their widespread use over the coming decade.

The next-generation seat, however, needs to take advantage of the integrated belt design and expand on its comfort and effectiveness. Borrowing the technology learned from racing, this seat is fitted with a *4-point* seatbelt. As with the 3-point system, the belts attach directly to the back and bottom of the seat, but now the safety-belt webbing rests uniformly on *both* shoulders. The even pressure from each strap means no more pulling and chafing across the collar bone, something that presently spells discomfort for millions of drivers and passengers on the road.

Built right into the seat, this 3-point belt system from Mercedes-Benz is the cutting edge of performance and passive safety for today's car market. Bruce Priebe - Twin Cities Section, Mercedes-Benz Club of America

Two seatbelt styles are possible. The first one merges two 3-point belts on your left and right side, crisscrossing at the chest and latching to the seat bottom on either side of your hips. The second, more traditional style, clasps over your abdomen just below the hips. The webbing passes over your thighs, firmly anchoring to the bottom of the seat on either side. The belts cannot ride up on you if the seat position changes. With the shoulder straps attached to the adjustable headrest, the correct position is maintained for your height.

Motorized shoulder straps make the second style of safety belts easier to put on. As you sit with your back flat against the seat, the press of a button brings the two straps automatically over your shoulders. It is then a simple task to clasp the harness together and give a slight tug on each belt to take the slack out. This buckle-up procedure is key to the success of the 4-point seatbelt package. When a safety measure is quick and easy to use, it is better received by the driving public. If it is comfortable, its continued use is assured.

As expected, the seat's framework is strong and rigid with the safety belts attached directly to the back and bottom. It won't break apart or pull loose from its mooring on the floor, even in a severe crash. The seat back still has several degrees of tilt to allow the most comfortable driving position for your body, but it does not recline completely flat. The upright position of the seat back and downward curve of the seat bottom will prevent you from sliding under the lap belt. The side bolsters are adjustable to fit the contours of your body, hugging your hips and back to reduce fatigue. A lumbar support maintains your posture and keeps you alert for driving, even hours at a time.

The ability of this seat to restrain you more effectively in an emergency is that extra measure of protection not found in most of today's cars. The problem of unwanted slack is minimized by locating the two locking devices for each strap under the headrest. With a strap over each shoulder, the energy of a crash is distributed more uniformly across the body and seat, further reducing the "spooling" effect. The addition of a manual "lock-in" lever for the 4-point harness also gives each person the option to be comfortably held in place while driving at high speed on the freeway. Such a lockable harness provides a psychological feeling of security. This device also deals realistically with the dangers of faster travel. During progressively harder braking, the driver and passengers can involuntarily lean forward several inches before the seatbelt engages. If a crash were still to occur, this is several inches of vital lifesaving distance lost. The lock-in feature would solve this problem for those using it. For those who do not, the clamping device right next to the person's shoulders would still provide improved restraint over today's seatbelts. A more g-force sensitive system could also limit such forward movement. This might include improvements as simple as all belts locking when the brakes are applied, greater sensitivity in the inertial-reel lock on the seatbelt or some type of electronic sensor to restrict belt movement. The cost of such innovations is contingent on whether you are driving an econo-box or luxury sedan.

The 4-point seatbelt system gives more comprehensive crash protection. In side impacts, the extra shoulder strap limits movement. During a rollover, the belt lower on the hips and dual shoulder straps hold you in place better, especially if the lock-in lever is used. For frontal crashes, the two shoulder straps more uniformly distribute the energy of impact. In a high-speed crash, the force of being snapped forward into a loose belt and the subsequent spooling can cause head, neck and body injuries in and of themselves. The snugger, 4-point belts reduce this possibility further. For even greater crash protection, the width of the belt webbing could be increased from the standard two inches to two and a half or a full-race, three inches wide. While such wider harnesses might only find their way into squad cars, sport sedans and exotics, a new level of evasive and lifesaving capability would be achieved if the average car of tomorrow used this kind of seatbelt system.

Cost-effectiveness is achieved by standardizing the seat frame using common materials like steel, aluminum and fiberglass. The 4-point design eliminates the need to manufacture right and left seats as in present-day 3-point systems. With the exception of the motorized belts, an economy-car seat could be manually adjustable, while a more upmarket seat would be electric. Whether covered in vinyl, cloth or Connolly leather, the seat and belt system would be identical from vehicle to vehicle in a manufacturer's line-up. Such uniformity would lower the price from thousands to hundreds of dollars. And this is only for front seats where the seat back is not attached to the roll-cage. The lighter, cheaper rear seats can be anchored directly to the vehicle's frame, again helping to hold costs down. A modest-priced vehicle could have almost the same level of seat and safety-belt protection as an ultra-expensive sports sedan. A fuller, more comprehensive level of improved safety would reach all passenger vehicles.

If this seat sounds at all familiar, it is the same basic design pioneered by Liberty Mutual Insurance for its Survival Car II back in the early 1960s—now updated and revised with improved technology. This modern version meets all the challenges of tomorrow's driving. In ergonomic terms, it does not restrict visibility. There are no belts attached to the doorpost to block your view. The headrest is directly behind, out of the way. For lane changes and looking to the rear, the view is unobstructed and clear. At low or high speed, drivability and comfort are first-rate. In all types of emergency maneuvers and crashes, you are held firmly in place.

Even with all these improvements in place, we still must address the question, what type of accidents will be encountered as we begin to drive faster on the open road? When a crash does occur, the outward appearance of the vehicle body would still look twisted and mangled. The passenger compartment, however, would remain intact. Belted drivers and passengers would walk away from all but the most severe, high-speed crashes. The three-quarters of accidents that happen under 40 mph would cause property damage, but little or no injury. On the freeway, where traffic moves in the same direction and roadside obstacles are shielded with guardrails or other energy-absorbing barriers, single-vehicle accidents or major pile-ups would cause no injury or loss of life, even at speeds above 75 mph. An 80 mph automobile crashing into another stopped car on the freeway has the same energy as the two colliding head-on at 40 mph.

With roll-cage and integrated seatbelt protection available in more and more vehicles, most collisions with life-threatening consequences would involve pedestrians or side and head-on crashes along rural highways. The human body and high-speed travel do pose limits on safety design, however. The exterior of a vehicle can be made only so "friendly" to those struck while walking or biking down the street. The vehicle's interior would give passengers a fighting chance to survive a solid-barrier crash (a tree or another vehicle head-on) in the range of 40 to 50 mph. Eventually, this might be extended slightly beyond 50 mph for some vehicles as roll-cages are further improved to deform progressively outward and away from the passengers in an extremely severe collision. The majority of crashes on tomorrow's road need not result in death and suffering—as long as passengers are wearing their seatbelts inside of a roll-cage-equipped vehicle.

So why all this fuss over crash protection? The obvious answer has been staring us in the face for years: This battle has been waged far more successfully on the racetracks of America than on public roads. Those who deal openly with the high-speed dangers of racing naturally take the precautionary measures necessary to make high-speed racing safer. Their willingness to seek solutions also speaks to the love and respect designers and drivers have for each other and the cars they build to race. Over the last thirty years, fatal race-car crashes have become increasingly rare while speeds on the track have risen dramatically. The same must be done for vehicles on public roads—especially if speeds are going to increase, as they most certainly will.

Most automotive manufacturers beam with pride that they support racing teams in order to bring the latest innovations from the track to the road. In terms of performance safety, that is quite true. Better tires, suspension and engine technology have made their way from the track to the road. But there have been countless other passive safety improvements in racing that have not made their way to public roads in the same period.

Safety takes on a new meaning for motorcycles and 18-wheelers, as well, when higher speeds are factored in. The average motorcycle of tomorrow will expand on technology already on the road. The fact that the best superbikes on today's market have reached a certain plateau of outstanding performance capability is undeniable. In terms of acceleration, handling and top speed, few vehicles on the road can touch them. Some improvements like antilock brakes and stickier tires will find their way onto more bikes over the next decade. Integrating "spill bars" into the aerodynamic fairing around the rider's legs will give some measure of protection if the bike goes down. However, crash protection for motorcycles is almost nonexistent, and a helmet law in exchange for faster riding is window dressing at best. Accident avoidance through proper training and performance safety design is the *only* way to reduce motorcycle injuries and fatalities as speed increases. About 2,000 people die each year in motorcycle crashes. Since passive safety is the weak link in motorcycle design, other precautions must be taken to limit the number of crashes for those who ride.

To move tomorrow's interstate commerce more efficiently, 18-wheel semis can take advantage of several innovations to increase speed, improve safety and lower operating costs. The first investment is in improved aerodynamics. Streamlining the exterior of the tractor and trailer helps to reduce wind resistance at higher speed, lowering fuel use and air pollution. Ultra-aerodynamic bodywork could raise the miles per gallon on the average rig from 6 mpg at 65 mph to 10 mpg at 80 mph, moving commerce faster at reduced cost. Safety would not have to suffer with these improvements, either. Rounding off the rig's front end would enhance the driver's visibility. Stability at high speed and better crosswind protection would be a by-product of this cleaner design. Enclosing the length of the rig with "windskirts" on either side would channel the air more smoothly around the sides and bottom. Reinforced with a separate metal subframe, these skirts would also act as "underride" protectors in a crash, restricting smaller vehicles from sliding underneath the trailer. Lower, more rounded bumpers would serve the same function. Covering the wheels in this fashion would also restrict the amount of spray thrown up by the tires in rain and snow, making it easier and safer for other drivers to see around these big rigs to pass. In high crosswinds, however, the windskirt itself could be raised, reducing the trailer's profile but leaving the protective cage underneath.

Greater use of aluminum and composites could further reduce weight, improving fuel mileage. Antilock brakes will continue to bring enhanced stopping and steering control, helping to slow the rig more quickly and reduce the possibility of "jackknifing." Radar-activated brakes are also coming onto the market, giving heavy trucks a small edge at stopping more quickly when a vehicle cuts in front or slams on its brakes. The radar unit is connected directly to the truck's brake system, monitoring the speed and closing distance of the vehicle ahead. If it gets too close too quickly, the brakes automatically apply, slowing down or stopping the rig.

If any of these heavy-truck innovations are to be brought to the American highway, there is another facet of trucking and truck design that must be addressed. That is the regulation of heavy trucks used for interstate commerce. No other area of safety related to transportation efficiency is in greater disarray. Many safety and law enforcement officials have an almost pathological hatred of heavy trucks, and rely more on preconceived notions and misinformation in determining safety policy than on sound engineering criteria. Each state operates its own weigh-station/enforcement fiefdom, ranging from good to intolerable. But little, if any, reciprocity exists between the states to make trucking more efficient or safe.

Imagine if you were leaving on a cross-country trip and at the border of every state had to stop for an hour or more to have your vehicle weighed and inspected. Your previous stop would mean nothing, and many of these inspections would nit-pick for the tiniest flaw, charging you extra money or delaying you further until it was fixed. You would have a never-ending stream of forms, log entries and citations to fill out, many of them redundant or of little value. The number of hours on the road would be tallied up, keeping track of your time behind the wheel with a confusing system of hours per day and total per week. As an average motorist, you would probably be outraged by this regimen. Instead of driving you'd either stay home or take some other form of transportation. Truckers, however, have to deal with this reality every day.

The legacy of this tangletown of rules and regulations is an adversarial relationship between safety and transportation efficiency. Innovations like antilock brakes took decades to be required. Comprehensive underride protection has yet to be mandated. Mysteriously, the length of trucks has been allowed to grow from single- to double- and even to triple-long trailers, in spite of the fact that most truck drivers don't like operating these "road trains" on the highway. Safety experts "talk" about lowering heavy-truck fatalities by requiring tougher vehicle inspections, shorter hours of operation and 55 mph speed governors on all 18-wheelers—all in a country 3,000 miles long.

This is just a sample of the problems associated with trucking in America today and we all pay for it. The price of goods, taxes and insurance premiums are all adjusted to reflect the draconian, inefficient way we move trucks down the road. To resolve this problem, we must start treating interstate commerce as a *national* agenda, one that sets up *total* federal control of trucking for uniform regulation from Maine to California. This agenda must include such items as: Commonsense changes for weighing in, inspection and safety innovation; the creation of a federal license plate for all heavy trucks which prorates the fee based on miles driven; and transformation of the present glut of paperwork into a more user-friendly, paperless computer network that minimizes time off the road and falsification of data. Such reforms would help to put fleet carriers and independent owner/operators on a more profitable basis, ending excessive regulation by the government and abuses by a minority of truckers. Again, speed is the key to make this happen. There is absolutely no incentive at the present time to improve on interstate commerce and its regulation with the reduced speed limits for heavy trucks in many states. Legally permitting trucks to operate in the range of 70 to 80 mph on the rural interstate from coast-to-coast will quickly advance rule-making and safety innovation from talk to action.

Implementation of such changes could come through tying improvements directly to speed, allowing new trucks and those retrofitted with radar/antilock brakes and windskirt/underride protection to be driven faster than those that are not equipped. Stenciling a larger inspection sticker/bar code onto the tractor and trailer could limit the number of stops at tollbooths and weigh stations. Law enforcement and toll attendants would be able to "read" the correct weight and charge the right fee without stopping the truck, keeping average speeds higher. (Some toll roads are already using sensors to automatically record fees. Some inspection stations and a growing number of trucks have "prepass" transponders that allow rigs to skip the checkpoint.) A rigid, comprehensive mechanical inspection for each rig could take place every six months or so many miles. The attached sticker or bar code would confirm the inspection and be valid from coast-to-coast—no more guesswork on the part of law enforcement, no more unscheduled stops for truckers. Hours of operation could be *increased*, as long as a more rested truck driver is behind the wheel. Streamlining regulations and boosting average speeds could significantly increase the number of miles driven during the typical work day. Working together instead of fighting one another would dramatically raise the efficiency and safety of road commerce.

A more balanced attitude toward truck safety should work its way through law enforcement and the safety community. Irrational fear and anger often rule the debate on trucks and the accidents they cause. The mere thought of a massive, overweight truck careening out of control sends shivers up the spine of many drivers. When one does crash, the media generally give it plenty of coverage—as with an airline disaster—even though the vast majority of transportation deaths are caused by *passenger* vehicles. Approximately 30,000 people are killed annually in cars, minivans, pickups and sport utility vehicles. Tractor-trailer crashes claim about 5,000 lives. The statistic that gets bandied about to stir up outrage and paranoia is that most of these fatalities were in passenger vehicles. Only about 500 were the truckers themselves. These facts fail to show that three-out-of-four heavy-truck crashes are induced by drivers in passenger vehicles. Some measure of culpability rests with motorists who do not realize the extra space and braking distance required for tractor-trailers to operate safely.

How we analyze and interpret statistics must also come under closer scrutiny if we are going to improve truck safety and efficiency. The Insurance Institute for Highway Safety reports that one-quarter of all heavy trucks have some chassis defect serious enough to take them out of service. But a study done by the University of Michigan indicates less than 6 percent of truck accidents are caused by some electrical or mechanical problem. Certainly, the fact that 24 percent of all heavy-truck fatalities occur on the freeway should be cause for concern when setting out to permit tractor-trailers to operate at higher speed on the interstate system. A growing percentage of trucks are already moving down the freeway between 70 and 80 mph. It is highly unlikely that they will be slowing down.

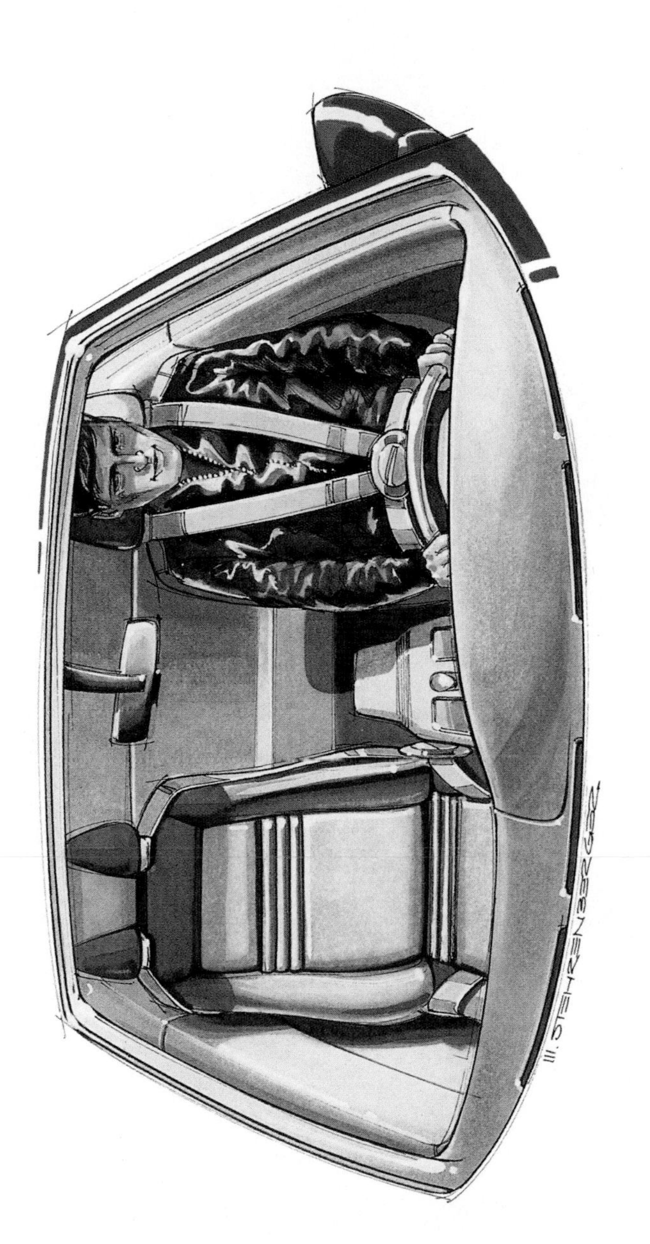

The 4-point integrated seatbelt system gives more comprehensive protection in all types of crashes and emergency situations. It is also easy to use and extremely comfortable—like wearing the vest of a fine suit. (Chapter Five).

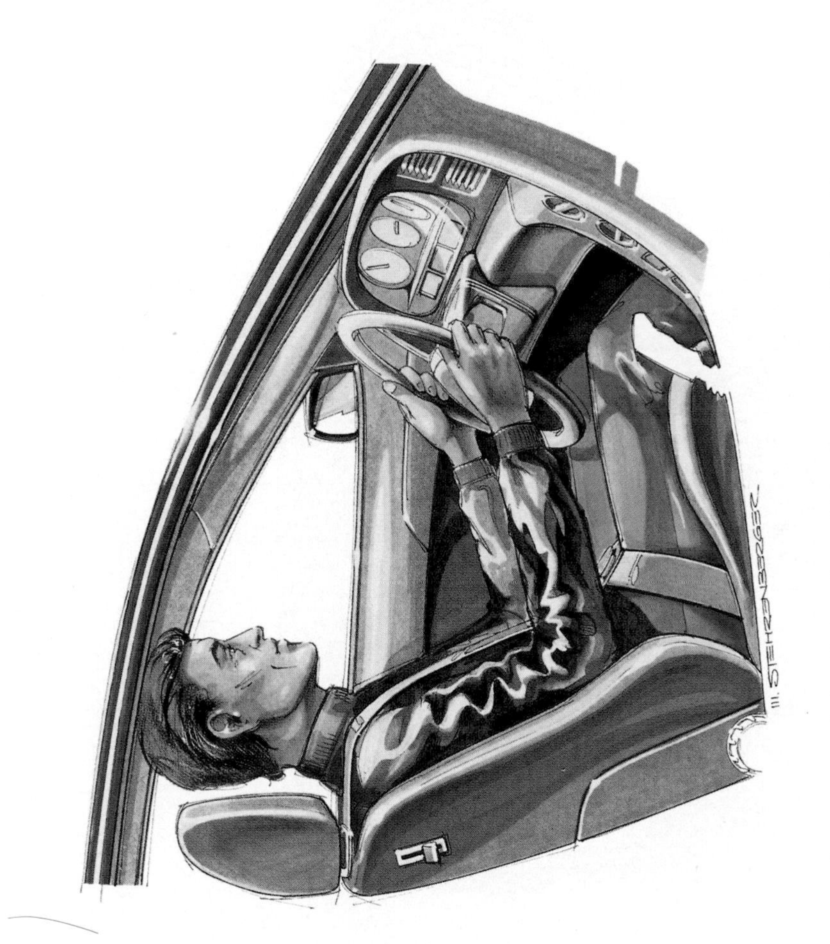

Seat position, comfort, ease and simplicity of reaching the controls… all this adds up to proper ergonomics— the foundation for safe driving at any speed. (Chapter Five).

Already close to the pinnacle of performance safety today, the superbike of tomorrow will subtly refine suspension, brakes and tires for improved high-speed handling. However, with an almost total lack of crash protection, a safer motorcycle will require improved education for both riders and drivers to reduce accidents. (Chapter Five).

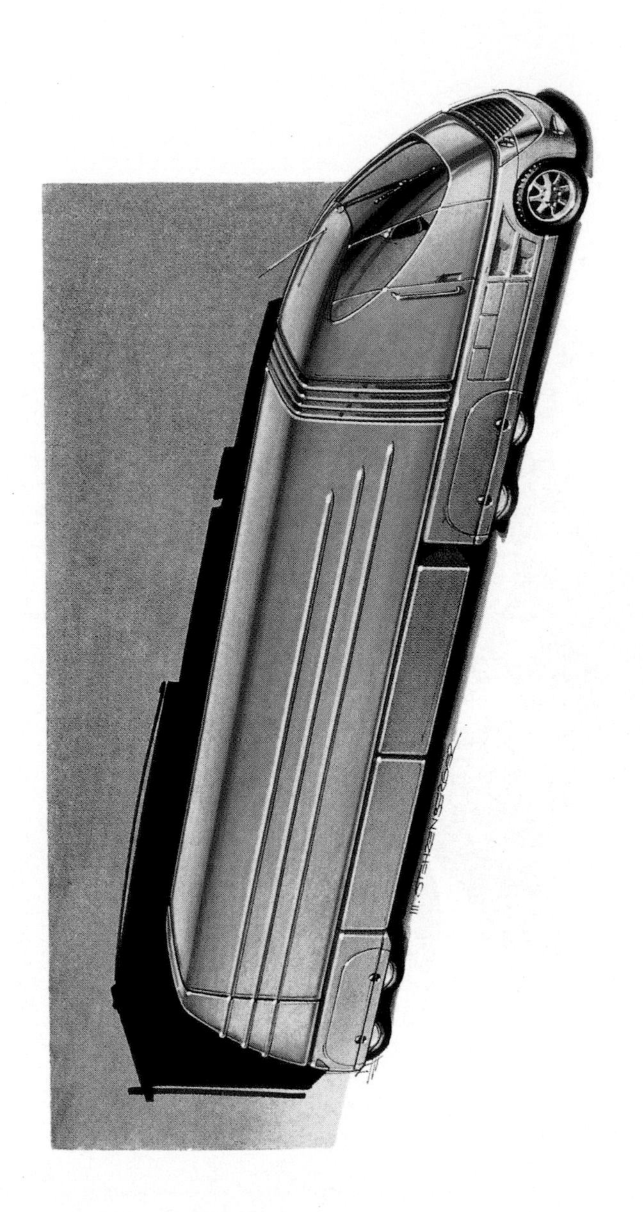

The 18-wheel tractor-trailer for the next century. Ultra-aerodynamics, 10+ mpg at 80 mph, antilock/radar-activated brakes, protective "underride" bumpers and windskirts. With proper national regulation, these big rigs would move the country's goods, faster and safer than ever before, and would no longer be feared by the average motorist on the road. (Chapter Five).

Interstate patrol car for 2000 and beyond. This 150 mph supercar has a 3 liter, 300 horsepower V-8 under the hood and a dash full of high-tech communications and monitoring equipment to oversee a safe flow of high-speed traffic on tomorrow's freeway. (Chapter Eight).

From the tuned exhaust to the wide-track, height-adjustable suspension, this squad car is ready for the rigors of daily high-speed driving on the beat. Rear-facing "arrow stick" above the rear window pulses to the left, right or on-off to inform drivers which direction to go or to stop in an emergency situation. (Chapter Eight).

ILLUSTRATIONS BY MARK STEHRENBERGER

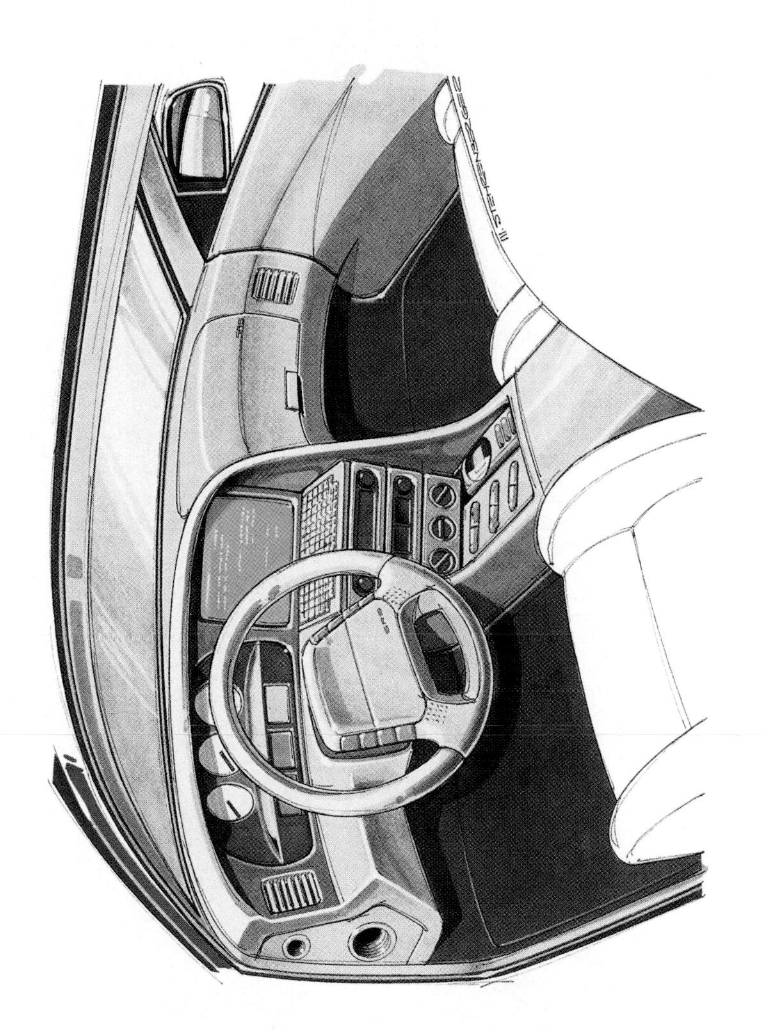

Interior of tomorrow's squad car. Built right into the dashboard and steering wheel, the radios, computer keyboard and important switches are at the fingertips of the officer in the driver's seat. Such a layout provides outstanding ergonomics for cruising the beat or driving flat out on an emergency, lifesaving run. (Chapter Eight).

Fiftieth anniversary Corvette and Lincoln Mark "X" on tomorrow's freeway with no speed limit. High-speed motoring enjoyment with outstanding safety. (Chapter Nine).

Dealing more openly with the speed of heavy trucks is a tremendous impetus for the safety innovations needed to save lives and *move* commerce.

As an adjunct to this discussion of tractor-trailer regulation, the mechanical inspection of passenger vehicles also becomes more important as speed increases. Once safer vehicles are put on the road, keeping them safe is another interrelated area of improved vehicle design. On today's modest-paced American highway, the need for mechanical inspection is not readily apparent. The dynamic stress on most passenger vehicles at a legal 65 to 75 mph is not enough to induce a significant number of accidents. Moving in the direction of speed limits above 75 mph, or no speed limit at all, could raise the percentage of accidents due to mechanical failure from single to double digits. The forces acting upon the frame, brakes, steering and suspension multiply exponentially as speed increases. A thorough, but flexible, program for overseeing the structural and mechanical integrity of passenger vehicles should be phased in state by state. Minimum standards could then be set for the load-bearing areas of the frame and suspension, preventing accidents due to excessive rust or damage. Brakes, steering components and tires must not suffer from severe wear or looseness that could cause loss of control. Seats, seatbelt mountings and the belt webbing itself should not be broken, torn or frayed. Fenders and door panels flapping in the breeze—in fact, anything that could threaten visibility or pedestrian safety—would have to be repaired. This would also be an excellent opportunity to spot-check emissions for vehicles that are potential "gross polluters."

Basing this type of safety inspection on adequate repair as opposed to mandated replacement makes the program more viable and cost-effective. Allowing a backyard mechanic to repair faulty brakes with good, used parts from a salvage yard could spell the difference between an inspection program that is implemented and one that is just talk. A loose, rusty fender fixed with a patch of welded or pop-riveted steel means vehicle safety could be extended across all economic levels. This would not be possible if a new fender had to be installed by a certified body shop in order to put the car back on the road. Such patchwork solutions are not perfect, but they do move forward the concept of increased speed with enhanced safety.

A safety inspection every other year would not only prevent accidents due to mechanical failure, but also enhance the psychological benefit to the driving public that comes from knowing that the vehicles on the road are "safe." The importance of this driver confidence should not be underestimated. The assurance that all vehicles have met a comprehensive minimum standard will take some of the uncertainty out of driving faster. Whether passing a sleek, new 18-wheeler, or being passed by an unrestored muscle car from the 1960s, drivers would feel more at ease than they do today. Presently, twenty-one states have some kind of inspection program. Almost half the Union is already monitoring vehicle maintenance. Over the next decade, that number will undoubtedly increase as we see an open use of speed on our best highways.

In the effort to improve the design of new vehicles, crash testing plays an essential role. At present, however, we have turned this testing into a make-work cottage industry that does little to bring improved safety innovation to the driving public. Among the automobile manufacturers, insurance industry and government test facilities, *thousands* of vehicles are crashed annually into laboratory test barriers worldwide. One would think, with so many cars and trucks being sacrificed to accident testing, that no one would ever die on the road again. The fact that we have known for more than forty years how to build a "safe" car, along with the reality that it only takes about a dozen or so test vehicles to prove it, gets lost in a modern-day crash-test barrage gone mad. This winds up costing the public billions of dollars in higher car prices, insurance premiums and taxes—while it takes decades to bring some of the simplest innovations to the road. Some improvements, like computer simulation, *have* helped to cut down on the number of vehicles smashed into test barriers, but rectifying the remaining problems will require a rewrite of the rules and regulations governing crash performance.

The federal government mandates several crash-test requirements that fall under the following Federal Motor Vehicle Safety Standards (FMVSS):

FMVSS 204 -	Steering Control Rearward Displacement. 30 mph into a flat barrier.
FMVSS 208 -	30 mph Front and Angle Barrier Collision, or "Hyge" sled test. (With and without seatbelts for airbag-equipped vehicles)
FMVSS 212 -	Windshield Retention. 30 mph into a flat barrier.
FMVSS 213 -	Child Restraint. For restraints built into cars, 30 mph.
FMVSS 214 -	Side Impact. 33.5 mph lateral by moving flat barrier.
FMVSS 219 -	Windshield Zone Intrusion. 30 mph into front end with flat barrier.
FMVSS 301 -	Fuel System Integrity. 20 mph side, 30 mph rear by moving barrier, 30 mph front flat and angle barrier.

All of these tests use a fixed or moving barrier that the vehicle strikes, or is struck by. This smooth, flat structure permits the most uniform, progressive destruction of a vehicle's front, side or rear end. Testing with such a barrier is, however, not particularly representative of the types of crashes that happen in the real world. (Nor is the "Hyge" sled which shoots a car backward at 30 mph for 1/8 of a second to simulate a crash.)

The type of tests used and the way the results are reported must be upgraded to provide superior passenger protection. The crash barrier itself needs to better simulate the front end of the standard automobile and its ability to deform in a collision. For head-on crashes, the test vehicle should strike this barrier in the way most front-end collisions occur in the real world on two-lane roads: left-front headlight to left-front headlight. This kind of "offset" crash puts increased strain on the driver's side, allowing greater intrusion of the vehicle's unibody into the passenger compartment. Requiring the speed of these tests to be raised from 30 to 35 mph (most manufacturers already test at 35 mph) would quickly magnify the need for such improvements as reinforcement to the unibody subframe, more interior padding on the doors and better seatbelt design.

One additional test must also be added to this crash-test regimen. Today's anemic rollover standard is woefully inadequate to provide proper passenger protection. Instead of requiring a vehicle to support one-and-a-half times its weight on the roof, a more formal dynamic test should be employed. Launching the vehicle from a tilted platform at 35 mph would better simulate the type of rollover crashes that happens on the road today. Such testing would require the strengthening of the roof pillars on most vehicles in order to pass. Finally, basic roll-cage design would come to all new vehicles sold in the United States.

If we are going to get serious about reducing the number of people killed and injured in motor vehicles, we must not mandate auto-safety rule-making in specifics, requiring endless FMVSS numbers for this and that. The federal government must stop micromanaging safety design and, more generally, demand auto safety criteria that provide injury-free protection in front-, side-, rear-end and rollover crashes. However manufacturers accomplish this is their business. A new vehicle either passes or fails the battery of tests without mountains of paperwork or governmental bureaucracy. A staff of government, insurance and industry analysts would preside over this open testing done by the manufacturers. Simple, straight-forward and cost-effective, it would be undertaken only for new vehicles entering the marketplace. At the time of a new car or truck's release, this comprehensive series of crash tests would determine whether the designed-in safety features worked properly. If the model continues on for several years with only minor cosmetic changes, further testing is a waste of time.

Such testing must also include the "seatbelt imperative." No amount of padding, structural rigidity or airbags can replace the absolute need for the use of seatbelts if the occupants are to survive *all* types of crashes. Unfortunately, the federal government is now requiring unbelted crash testing for vehicles equipped with airbags. The goal is to monitor this *supplemental* crash protection for its effectiveness as a *primary* restraint. Such a test sets a dangerous precedent. Granted, for the 40 percent of Americans who don't wear seatbelts, this is a serious safety concern, but airbags will never fill this gap. For accident avoidance or crash protection to be effective at all, safe vehicle design must rest on the foundation of *everyone* being buckled up. To illustrate this as definitively as possible, Mercedes-Benz currently builds the safest cars in the world in terms of performance and passive safety. However, if seatbelts are not worn, it negates *all* of those safety features—as Princess Diana and her bodyguard, Trevor Rhys-Jones, tragically proved.

After having passed this series of new crash tests, the vehicle should have prominently displayed on its purchase sticker two numbers. A lower one for zero injury (at a minimum speed of 25 mph) and an upper number for survivability (at a minimum speed of 35 mph). These numbers could have four arrows around it to indicate that the vehicle has attained front-, side-, rear-end and rollover crash effectiveness at these speeds. More safety-minded marquees like Mercedes-Benz and Volvo could strive for even higher numbers. A vehicle stamped "40," "45" or even "50 mph" would send a clear message of superior crash protection to prospective buyers. This would end the guesswork by consumers trying to understand the confusing five-star system the federal government now uses. This ridiculous rating scheme only measures injuries to the head and upper body. Vehicles can receive a perfect five-star score while the legs and feet could be mangled to a pulp. A new mile-per-hour, zero-injury/survivability system, along with broader rule-making and crash-test requirements, should help eliminate the gaps and excesses that presently exist in mandating safe vehicle design. For tomorrow's road, a comprehensive, minimum standard of passive safety would finally be achieved.

All this talk on vehicle safety can be reduced to one question: How effective will it be? Can we expect a dramatic reduction in the number of people killed and injured over the next two decades if certain innovations find their way into all vehicles? This ultimately depends on how well we integrate the human factor into the automobiles of tomorrow. Already we can claim to have achieved on today's market "The Best Safety That Money Can Buy." There is a smorgasbord of performance and passive safety to chose from: four-wheel drive, speed-sensitive steering, "active" suspensions that sense the bumps and curves in the road. Side-impact airbags are popping up in more and more models. Mercedes-Benz is selling radar-controlled cruise control and computer-monitored steering to prevent spinouts. The list goes on and on. But the fact remains that the safest car in the world can be made deadly with a dangerous driver at the wheel. Success in saving lives and reducing injuries will only come through the recognition that *you* are in charge when in the driver's seat. Safety can be built around you, but technological improvements cannot replace your common sense and good judgment. Unfortunately, the American safety community has continued to look at the automobile as an inanimate object, a thing that can be made safer simply by cramming gizmos and gadgets inside to improve crash protection. The human element has been incidental to the process. The overriding feeling has been that accidents are going to happen anyway, so we must protect people from themselves. The individual behind the wheel is looked upon as too stupid to handle the task of driving. However, since the automobile cannot drive itself, the laws and rules of the road must then reflect the lowest common denominator.

This attitude toward safety held by the government, insurance industry and consumer advocates, has come at tremendous expense to the American people. We continue to pay a portion of our taxes to support a federal bureaucracy that strives to mandate cleaner, safer vehicles. A percentage of our insurance premiums effectively does the same thing, funding the Insurance Institute for Highway Safety to lobby for changes (through companies like State Farm, et al). The final insult is paying for these mandates, regulations and legal battles a third time through the inflated purchase price of a new car. All the while, the annual highway death toll continues to hover around 40,000—another expense we pay for in lost productivity, medical costs and increased travel time. While the automotive manufacturers must also share a portion of the guilt, they have now been backed into the wall by this monolithic safety "empire"—

especially in the case of the insurance industry—that forces them to fight some mandates, like airbags, all the way to the Supreme Court to set the legal precedent. (And given the growing number of unintended airbag deaths, this now looks like a very prudent thing to have done.) The reason we have failed to save lives on the road is that the safety movement took a twenty-year detour in its struggle to mandate airbags while the structural integrity of the passenger compartment was basically allowed to languish. Before an airbag can be reasonably effective it must be technically perfected. It must also be installed in a vehicle with a rigid roll-cage where seatbelts are being worn. Safety proponents naively felt that giving everyone an airbag and a speeding ticket would result in more lives saved. This has not been the case. They have failed to work for the three things that could dramatically reduce the number of people killed and injured on the road: the use of seatbelts by 99 percent of the population, the phase-in of roll-cage safety for all vehicles, and the expectation that the average motorist will behave as a responsible, reasonable and prudent adult.

The Fast-and-Safe philosophy, however, can move us more quickly in that direction. A demanding high-speed driving environment will make the motoring public and auto manufacturers sit up and take notice. The marketplace and the needs of the driver will motivate safety innovation more than will bureaucratic mandates. With the speed of freeway traffic moving in the range of 80 to 100 mph—legally—many devices like underride protection for heavy trucks and roll-cages for passenger vehicles would be on the road in a much shorter time frame. Simply declaring majority rule would necessitate a 80 mph speed limit on the entire interstate freeway system today. Further, tying any future increases in speed limits to seatbelt use would do more for highway safety than the 10 million speeding tickets written last year. Since safe vehicle design is contingent on buckling up anyway, such a trade-off should be more effective and probably more palatable to the American people than a seatbelt law that does not include an increase in speed.

The vast majority of motorists do not want to get into a crash, hurting themselves or anyone else. With this as the basis for safety design, the performance characteristics of a vehicle should enhance the enjoyment of driving through better response from the person behind the wheel. In an emergency, the *driver* is in command and avoids the accident through braking and steering input. If that is not enough, the driver has still helped to minimize the impending collision by slowing down and steering away. When the crash occurs, the vehicle absorbs as much of the impact as possible, with the roll-cage/seatbelt combination working to prevent injuries. Though the unibody and running gear do much of the work, it is the person behind the wheel who is in charge. The driver is a vital, active participant, not some hapless victim witlessly bashing into someone else. With the Fast-and-Safe philosophy, the person behind the wheel is the missing link for saving lives and reducing injuries. Vehicle design, however, plays no less an important role.

This symbiotic relationship between driver and vehicle is not without its problems. Some research indicates that all our efforts to bring safer cars and trucks to the marketplace might not be helping to reduce significantly the number of people killed and injured. The thesis of this research is that the benefits of safer vehicles are "offset" by some motorists taking greater risks behind the wheel. Since only a small percentage of the 170 million licensed drivers in America get into serious accidents, a tiny increase in aggressive behavior could be negating the effect of safety improvements in vehicles.

This theory of offsetting behavior was put forth by Sam Peltzman in the mid-1970s. His research indicated that driver behavior may change in response to mandated automotive safety improvements, canceling out their effectiveness. In simple terms, it's like switching from a typewriter to a word processor. On the typewriter, pushing the wrong key means white-out or retyping the page. On the computer a mistake is a minor problem: back space and type over. The level of risk for a typo drops almost to nothing. You pay less attention, taking more chances by typing faster. In a safer automobile, the perception of risk appears lower. Some motorists drive more aggressively and get into accidents. Unfortunately, it is not possible to back up and drive over.

More recently, research in offsetting behavior has been undertaken in separate investigations by George Hoffer of Virginia Commonwealth University and Robert Chirinko of Emory University.

Using relative personal injury and absolute collision insurance claims, the Hoffer study compares vehicles with or without airbags as they were being phased in during the early 1990s. The findings suggest that airbag-equipped vehicles had a higher number of relative injury and collision loss claims than vehicles without. It contends that "at risk" drivers, whether out of individual driving style, location or high mileage might be reducing airbag effectiveness through increased exposure and aggressive driving habits.

The Chirinko research takes a broader overview of seatbelts, and airbags, as well as the 55 mph national speed limit before it was repealed. Again, this study corroborates that motorists will more than likely modify their driving habits with a change in their safety environment. This response will probably lessen the impact of safety regulation on reducing total motor-vehicle fatalities. With safer vehicle design, we could expect a reduction in fatalities inside of vehicles but an increase in pedestrian and bicycle deaths through more aggressive driving. Further research suggested that repealing the 55 mph speed limit (slightly increased risk) and requiring airbags in all vehicles (somewhat reduced risk) would only lower total fatalities by 2.7 percent, not the 33 percent estimated by safety engineers. Given that over the last decade fewer than 2,000 lives have been saved by airbags in more than a quarter-million deployments, this research appears to be pretty much on the mark. It is unlikely that airbags could survive a cost-benefit analysis. American car buyers have spent *billions* for a "safety" device with little overall value or effectiveness.

It is important to note that offsetting behavior does not mean we become crazed maniacs speeding down the road in our new, "safer" cars, willfully trying to have an accident. The effect is much more subtle. It is the person who decides to drive on a rain-slick or snowy evening, the individual who has just one more beer before heading home, or the motorist who rushes the yellow light. After all, their new cars have antilock brakes and traction control. If all else fails there's the airbag. In most cases, these drivers reach their destinations without incident. But if, out of the millions of motorists who take these and other chances, a few of them don't make it, the margin of safety is diminished even though "safer" vehicles are on the road.

Quantifying such behavior doesn't mean we are going to abandon the notion of safe vehicle design and reintroduce the '61 Corvair with all kinds of spearlike switches on the dash. These studies don't recommend giving up on safety improvements, either. They point out that human nature combined with the automobile is an extremely complex relationship to analyze and understand. Offsetting behavior becomes one more factor to consider when striving to build safer vehicles.

Closer scrutiny of advertising might help minimize the offsetting effect. Striking a balance between the media hype used to sell safety innovations and proper information for the buyers about their limitations is a challenge. This relentless sales blitz is probably having a certain effect on people's perception of what safety options can or cannot do. The omnipresent twenty-point type: NOW WITH DUAL AIRBAGS! clearly overshadows the eight-point reminder to always wear your seatbelt buried somewhere on the page or screen. This does little to inform the public about the advantages and disadvantages of both safety devices. Antilock brakes have the same problem. A quick blurb that a vehicle has them doesn't begin to explain their complexity. Misconception and conjecture can follow.

Education of the motorist should not stop with more prudent advertising. A better understanding of the overall dynamics of accident avoidance and crash protection could help to reduce the false sense of security created by safer vehicles. Knowing how your automobile can be expected to react in an emergency situation or collision might modify your behavior in a more positive direction.

In a nonscientific, anecdotal way, the twelve-year research for this book helped to modify my attitude towards driving and safety. Learning about the idiosyncrasies of antilock brakes and airbags taught me not to assume too much. Knowledge of roadside safety and accident outcome caused me to drive differently than before. My perception of risk changed. Dangers previously unseen were now exposed, and the limitations of road, vehicle, and driver better understood. I wasn't lulled into thinking I was immune from any crash nor did I become paranoid about ending up as a statistic. My education on highway and auto safety simply informed me about the broad scope of the issue. I now drive slower under some circumstances, faster under others. Such knowledge should also help other motorists to deal better with the false sense of security that can be generated by improved safety features.

More attention to the economics of offsetting behavior could bring better, all-around vehicle safety to the American road. The trend towards high-tech gizmos and gadgets to save lives is driving up the purchase price and complexity of the average automobile. Since 1981, the average new-vehicle purchase price has skyrocketed from $8,850 to just over $20,000 in 1998. A growing number of consumers are beginning to ask with a certain indignation, "Why does my new car cost $20,000?" The answer lies in a complex set of rules and regulations that mandate expensive safety and pollution control equipment. Most of us never have to use all the safety measures built into our cars. They simply cost us extra money and aren't even guaranteed to save the life of someone who is at risk.

We could do a better job by mandating *low*-tech safety. This is where federal regulation started out in the mid-1960s. The initial mandates for padded dashes, collapsible steering columns and seatbelts brought some of the most cost-effective, beneficial changes to the automobile, providing the best return on safety for the investment. Most of us never had to use these devices, but the few dollars spent per vehicle was not excessive. Though most auto executives and automotive enthusiasts would be loathe to admit it, these safety improvements instigated by Ralph Nader's initial foray into improved passenger protection have helped to save many lives and to reduce by thousands the number of injuries over the last three decades. Today, however, excessive safety regulation is adding $2,000-$3,000 to the purchase price of the average car. This is quite an expense, especially when the total number of fatalities per year has not dropped noticeably. If, instead, we required a stout roll-cage, a good seatbelt system and solid running gear, we would probably do more to save lives than all the present regulations combined. With regards to offsetting behavior, the roll-cage would be hidden from view, and its effectiveness not readily apparent. There would be no doubt that seatbelts would have to be worn to achieve any lifesaving potential. The basic performance level of the chassis would give positive feedback to the driver but nothing more. Advertising might be keyed to "the joys of driving" instead of including overt references to safety features. There would be no antilock brakes, traction control or airbags, no computer to help you stop, no bag popping out of the dash to save or take your life. Eliminating these features could lower the price of a basic economy car to less than $10,000. The average family sedan might range somewhere between $14,000 and $16,000, not the present-day $18,000 to $20,000. Millions more Americans could now afford to buy a new car. This turnover would put a larger number of drivers in all-around safer vehicles than are presently on the road today. There would still be a cornucopia of lifesaving options to chose from, and luxury sedans, sports cars and exotics would continue to push the leading edge of performance and passive safety. Only the minimum standard of mandated safety would change to better accommodate driver behavior within the framework of more cost-effective vehicle design.

In the coming years, safety advocates, law enforcement officials and the motoring public will also have to acknowledge that a $60,000 Mercedes-Benz on the freeway at 120 mph will provide a more consistent level of improved road holding and crash protection than a $10,000 Ford Escort at 90 mph. Such a "pecking order" in performance and passive ability should be *encouraged*. In fact, it is vital to the safety/speed relationship that the public tolerate "safer" cars being driven faster so motorists will want to "step up" and pay for innovations that will increase speed while lowering the chance of injury and death. Dealing openly with our use of speed will get us to this point that much quicker.

We might even be better served by letting the marketplace and driving environment, instead of the federal government, steer innovations. The most desirable, sought-after cars still come from Europe. In large part, this is due to a philosophy that regards driving as a serious, adult activity. Motoring with exuberance (i.e. high speed) is an exhilarating experience to be enjoyed but not taken lightly. The corresponding road system takes this attitude into account, with speed limits above 80 mph or no speed limit at all throughout much of the Continent. This driving environment has created the fertile ground necessary to produce the best all-around performance safety related to vehicle design. The braking, suspension and steering for most European marquees are still a cut above the rest on the world market. Over the last two decades, the Japanese have very skillfully copied this attitude and design. Such mimicry has helped them to sell successfully in the American market, bringing into this country, until recently, superior automobiles which emulate those of Europe at a reduced price. This has left the domestic manufacturers trailing at a close third. Such international competition *has* helped to improve U.S. vehicle design. Unfortunately, the lack of a dynamic driving environment has prevented America from leading the world market in overall design innovation. Low speed limits and the corresponding low performance expectations for such a road system have taken their toll on U.S. auto manufacturers as consumers seek the real and perceived superiority of foreign products.

Here is where the creation of a high-speed interstate freeway system has the potential to bring advanced safety design to the road, while propelling American automobiles to the top of the world market. Instead of being lulled into complacency by modest speed limits, the country would be shocked out of its lethargy through faster driving. Manufacturers would scramble to meet the safety needs of consumers. The real and perceived dangers of speed might help to "offset" the offsetting behavior of motorists, getting them to take precautionary measures they otherwise would not. It is important that this increase in speed be dramatic. We shouldn't whittle away at it in 5 mph increments. Raising speed limits from 65 to 80 mph on rural freeways in the East and abolishing them west of the Mississippi would leave no doubt in anyone's mind about the need for improved safety. Vehicle design would quickly be modified to serve this more demanding environment. Debate on the existing gaps in safety would be met with more action and less talk because of speed. If the domestic auto manufacturers responded to this challenge first, they would be better able to move from a competitive third to a strong first place in the world market. The perception that American-made automobiles could handle the rigors of high-speed driving would aid sales abroad. Speed and the corresponding safety precautions would reinforce one another.

Whether we do anything or nothing with automotive safety over the next decade, one fact remains: Vehicles will never be as safe as their designed-in, lifesaving features. Human behavior will defeat some or most of their effectiveness. However, the triad of seatbelts, roll-cage and crush zones provides *the* vital safety innovations for reducing tomorrow's death toll. In this respect, the automotive manufacturers shoulder a disproportionate burden in our quest for saving lives. Striving to create a highway system best suited for their products will help to offset any extra expense in bringing safer vehicles to the road.

CHAPTER 6

Drive Right: The Secret of Fast-and-Safe Driving

Part One
Lane Obedience

There is a safety program that is all but absent on the American highway—a program that does not require a change in road or vehicle design. In fact, it is completely cost free and can be implemented on the road right now. It is *the* reason why the *Autobahn* is so fast and safe. It is a safety program that is long overdue for this country.

Drive Right, Except to Pass.

It is so simple, courteous and easy to do, one wonders why it has not been more strictly enforced on our highways. Unfortunately, the reason is quite clear. The Slow-is-Always-Safer movement has been obsessed with restricting speed for so long, that any policy that might improve speed efficiency has been looked upon as a bad thing. The failure was to neglect the many important safety considerations that Drive Right provides to ensure a smooth flow of traffic.

This was not always so. We actually started off in the right direction, following the *Autobahn's* lead as early as 1940 on the Pennsylvania Turnpike. With no speed limit for the first six months of its existence, each toll ticket had the rules of the road printed on it to help Americans unfamiliar with this new, high-speed driving environment. Prominently displayed on the stub was "Keep Right—Pass on Left Only." This helped to set a precedent on the American highway, and it worked as a rule of thumb. Lane obedience was pretty good before the 55 mph National Maximum Speed Limit was enacted, and a whole new class of dangerous and self-righteous speed enforcers slid over into the left passing lane to impede traffic. For those who remember driving during those pre-55 Glory Days, "Driving Right" meant a more relaxed, less aggressive time behind the wheel as slow and fast drivers worked together to create a freedom of the road that is just now returning to the American highway.

In the early 70s, my father let me drive with only a learner's permit from Peoria, Illinois to Minneapolis, Minnesota in our 1967 Chevrolet Impala. Bidding family friends goodbye, my dad began giving me real-world driving instructions to negotiate the interstate freeway system safely. Entering I-90, I was instructed to stay in the right lane and strictly observe the speed limit (then 70 mph). When coming up behind slower cars and trucks (many drivers, in those days, especially older ones, did not feel comfortable going 70), my father declared, "Signal first, make sure it's clear, then soup it up to 80, 85 and get around that guy." I was also told, in no uncertain terms, to get back into the right lane and slow down, making way for other, faster motorists on the road. With the onset of 55, this Drive Right attitude changed radically. Many states stopped posting minimum speed limits on the freeway. More importantly, they removed their Slower-Traffic-Keep-Right signs. New drivers were not educated, either in driver's training courses, or by their parents, in the Drive Right rule. The slow driver had now become King of the Road. True Believers in 55 began blocking traffic deliberately, raising the ire of other motorists who were simply trying to move down the highway at a speed which felt comfortable.

Across America today, lane obedience can vary wildly from road to road, state to state, and by time of day. Those fortunate enough to drive the Indiana Turnpike generally will find a Drive Right attitude that, at times, is as good as the *Autobahn's*, and many stretches of rural interstate do flow with a Slower-Traffic-Keep-Right reality that makes driving safer and more relaxed. On the other hand, those hapless victims of "Left-Lane Bandits" on Minnesota's freeways are doomed to an aggravating drive across my home state. No amount of signal blinking, headlight flashing, or even prayer will move these comatose drivers from the left lane. Sadly, a growing number of American drivers, both fast and slow, operate their vehicles with an insufferable arrogance, searching for any advantage they can gain over other motorists on the road. From the slow driver who is oblivious to—or intentionally tormenting—those who would like to drive faster, to the speed demon who impatiently cuts and thrusts through traffic, cooperation and courtesy have become casualties of the 55 mph speed limit. Just two words would permit everyone to coexist safely: Drive Right.

On the *Autobahn*, Drive Right (or *Rechts Fahren* as it is called in German) is more than just simple courtesy. It is the law. Passing on the right is illegal and motorists universally obey this rule of the road. Under normal driving conditions, no one ever overtakes on the right or blocks traffic by dawdling in the left lane. At first glance, Drive Right appears to be a rigid system to get slow drivers out of the way of faster ones so they can speed even quicker down the *Autobahn*. It is true that Drive Right enhances the speed efficiency of the German superhighway network. Not having to slow down for vehicles blocking the left lane means your average speed will be much higher on a long trip. But this law has many other safety considerations as well. With high speed the rule, not the exception, a structured, uniform traffic pattern is necessary to reduce the dangers of passing. Since many different types of vehicles use the *Autobahn* at varying rates of speed, overtaking one another in a consistent way becomes vital to safety. Always having the slowest traffic in the farthest right-hand lane eliminates passing on both sides. The speed of vehicles may vary, but overtaking occurs only on one side—the left. Motorists on the *Autobahn* can plan for this, recognizing that faster traffic *always* comes up from the left. Whether you are fast or slow, this law demands that motorists accommodate each other by moving into the right lane to let faster traffic pass.

Drive Right is an intrinsic rule of the road. Don't think of it as numbers on the speedometer, or something you don't have to do when no other traffic is present. Do it in-and-of itself. Never drive in the left lane unless you are actively passing some other vehicle, and then, only if you have the right-of-way. Don't force some faster driver already in the left lane to slow down, either. If you are there yourself and someone zooms up behind you, there are two international signals to show you that this faster driver would like to pass. The first is to turn on the left turn signal and, second—if you continue to block the way (when an opening comes in the right lane)—the driver behind you will flash the headlights on-off during the day and high-low at night. For the uninitiated, such as a slowpoke or Left-Lane Bandit, this may seem both arrogant and frustrating. But think about this another way. If you were waiting behind someone at a stoplight, and when the light turned green, the person in front of you decided just to sit there because he or she didn't feel like moving, you'd be pretty upset. It is the same as if you were to block someone's progress in the *left* lane of the *Autobahn* or any multilane highway. Don't impede someone else's progress. This will make driving more courteous and less aggravating—especially when you are in a hurry and the other drivers on the road move out of your way.

Drive Right keeps you sharp and alert while you're behind the wheel. It helps to break up monotony on the road. In the right lane for normal driving, you are relaxed, attentive, enjoying the free speed of the open road. When the time comes to overtake another, slower vehicle, this becomes a period of high concentration. On the *Autobahn* with no speed limit, this passing maneuver is executed in the following manner: A look in your rearview mirror to see if anyone is pulling up behind you to pass. Then, a series of quick glances between the side-view mirror and the road up ahead. This keeps you focused on your driving in relation both to what is in front and back of you on the freeway. If someone is coming up in the left lane at high speed, these quick looks in the side-view mirror also help you better gauge the closing distance, allowing you to decide if you should go or wait. Once it is safe to pass, you signal your intent, check your blind spot and ease out into the passing lane to overtake the slower vehicle. If you are already at top speed, you can simply maintain your head of steam, briskly passing by. If you are driving at a slower pace, you can accelerate as you pass, making the actual time in the left lane shorter. Treat your pass as a major event. Move smartly around and back over into the right lane as soon as the passed-vehicle's left front end is visible in your rearview mirror. Don't slow down in front of the vehicle you have overtaken, either.

Ease off the gas once you're well out in front of it. Don't annoy someone you've just passed by sitting right in front of the vehicle's bumper.

Common sense should not be a casualty while Driving Right on a no-speed-limit, *Autobahn*-style freeway. Drive Right efficiency also means Drive Right safety. Whether overtaking a single vehicle, or a series of vehicles bunched up in a row in the right lane, one has to adjust closing and overtaking speeds accordingly. On a wide-open *Autobahn* with sporadic traffic that is evenly spaced and well apart, you can pass vehicles with much greater speed than if they are spaced closer together. Great caution should be taken when overtaking a group of vehicles that are close together in the right lane, especially if it is a combination of cars and heavy trucks. Someone could dart out at any moment to pass. Here is a valuable *Autobahn* safety tip for high-speed driving: If you are closing in on a slower vehicle, or cresting a hill where visibility is at all limited, back off the accelerator and have your foot just on top of, but not touching the brake pedal, using aerodynamic drag to slow your vehicle as you slip past. This saves you valuable reaction time should you need to dive on the brakes and make an evasive maneuver. Once past, you can step on the gas and be on your way. Only a fool passes a string of vehicles at 100 mph when the group is flowing along at 60, because it is impossible to stop in time. If there is any doubt, **SLOW DOWN!** You can afford to take a couple of extra seconds to pass this clump of traffic. You can easily make up the time once you're further down the road on an open stretch. Remember, after you're done passing, look in the rearview mirror to see if anyone wants to pass you. Even when overtaking a long train of vehicles, you may notice a short break between two of them. If a faster car sneaks up behind you, be accommodating, slip over into that space in the traffic and let the faster driver pass. It's just one of the courtesies slow drivers give faster ones and vice versa.

After a time, you should begin to feel self-conscious when you are in the left lane. Being there for any reason other than actively passing slower vehicles should put you a bit on edge—looking for the first opportunity to move back into the more comfortable right lane. This edginess, however, can sometimes turn to anger and frustration when faster drivers creep up behind slower ones to "encourage" them to move over and let them pass. In this respect, the *Autobahn* is not a perfect road. Some overly aggressive drivers have, in the past, intimidated slower drivers sitting in the left lane by riding their rear bumpers. But instead of trying to slow motorists down, German safety officials imposed a law in 1990 to discourage following too close, and the *Autobahn* police are not shy about handing out tickets to drivers who draft behind vehicles while not staying at least two seconds behind. This dangerous behavior generally happens when a slow driver dilly-dallies while passing. Most faster motorists will give you plenty of room if you pass smartly and in a forthright manner.

Drive Right and the lack of a speed limit complement one another well on the *Autobahn*. Drivers pay less attention to their speedometers and more to their driving and the road. No longer does one have to worry about transgressing some arbitrary, posted speed limit. Passing can be done attentively and authoritatively, using the accelerator as needed to move safely past other vehicles. Speed simply becomes an enjoyable tool to use to reach your destination as quickly as you feel comfortable. Adherence to the Drive Right law gives all drivers more confidence. Instead of blocking traffic, slow drivers are on the lookout for faster vehicles when they move out into the left lane to pass. The highballers on the *Autobahn* are watching for slower traffic pulling out. This helps to set up a symbiotic relationship that works extremely well.

I personally took part in this symbiosis many times on the *Autobahn*. Once, while driving south on the A-3 near Passau, Germany, my car was wound up to its maximum of 110 mph. Off in the distance was a slower Volkswagen Golf, traveling in the right lane as I was. As is habit on the *Autobahn*, I looked back in my side-view mirror—*way* back. Coming up behind me and already in the left lane was a growing white speck of a car that was obviously moving a great deal faster than I was. In such instances, the human ability to sense both speed and movement is amazing. Computing almost instantaneously the closing distance of these three vehicles, I realized I had just enough room to overtake the Golf while not forcing the white car behind me to slow down. If there had not been enough time, ***I*** was the one obligated to slow down and wait for the faster car to pass. Signaling my intent, I eased out into the left lane as the white car steadily gained on me and I on the Golf. At this rate of speed, a smooth passing maneuver uses a fair amount of road. Drawing ever closer to the Volkswagen, I knew if the white car had to slow down on my account, the driver would be irate with me for trying to get by the Golf first. This is not arrogance at play here, just the frustration associated with the long time it takes to build up your cruising speed once again. I shot by the Golf, tucking into the right lane as soon as I was safely past. Close on my heels and now clearly visible, a pearl-white BMW 700 series sedan zoomed by me at about 130 mph. I caught a glimpse of a faint smile on the face of the driver, a woman about forty years old, as she pressed on down the *Autobahn* and over into the right lane herself. None of us had to slow down to accommodate the other. We were all able to maintain our respective top speeds safely because of our attentive driving and rigid obedience to the Drive Right law.

To the True Believer in the 55 mph speed limit, or drivers who do not have a taste for speed, this incident on the *Autobahn* seems, perhaps, outrageous and absurd. It is one of the more extreme examples of the present differences between the German freeway system and our interstate; but, it clearly shows the different attitude at work to make speed as safe as possible, with drivers working together to ensure that this happens. Drive Right means knowing yourself as a driver. Are you fast, modest paced or slow? It means being accommodating to those around you.

When I drive around town, I am, by most people's standards, a slow driver, driving within 5 mph of the posted limit. I make an effort as a slow driver to Drive Right whenever possible, letting faster traffic slip by on my left. I can drive more relaxed because no one is riding my bumper (I will speed up or slow down if a Left-Lane Bandit attaches to my side), and faster traffic can uneventfully continue on its way, unimpeded by me. Unfortunately, once I am out on the open road and my pace picks up considerably, I notice many other drivers do not have a similar Drive Right attitude. Millions of Americans, like me, are trapped behind these inconsiderate motorists who will not yield the right-of-way.

The American highway safety establishment steers away from dealing with the Drive Right question by citing various studies that show the potentially deadly effects of speed variance—slower traffic being passed by faster traffic. Research undertaken by West & Dunn and Ezra Heier *does* show that variation of speed between vehicles is a problem to be reckoned with. In essence, these studies indicate that a uniform flow or "pace" of traffic, whether it is moving at 50, 75 or 100 mph is "safer" than one lane moving at 50 and the other at 75 mph. This is due to the possibility of slower vehicles pulling out in front of faster ones, creating a greater accident risk. Many traffic safety engineers are firm believers that uniform flow of traffic is the only way to enhance highway safety. What they fail to see is that Drive Right, Except to Pass can effectively deal with this problem.

Out in the real world, on the American highway, uniform pace is an impossibility. It does not exist. It is a theory that works well for computer models and looks good on symposium papers but has little to do with road safety in reality. The infinite variables that exist on our highways mean speed variance will always be a part of driving: Heavy trucks climbing steep grades while passenger cars whiz by. Older drivers who feel comfortable at 60 mph while some young hot rodder streaks by at 80 mph. *That* is the real world.

Dealing with speed variance in reality, and not theory, is the route to greater safety. These studies further conclude that a variance in speed under 10 mph appears to have no apparent effect on safety. Yet the *Autobahn* shows a dramatic variance in speed while achieving an equal or better safety record than our interstate. Subtracting the average speed of 80 mph from the 85th percentile speed of 96 mph on the *Autobahn* reveals a variation of 16 mph. At the opposite ends of the spectrum, the 15th percentile speed for heavy trucks on the *Autobahn* hovers around 55 mph while the 85th percentile speed for passenger cars is in the range of 100 mph. That is a 45 mph variance in speed! Germany's religious obedience to the Drive Right law yields a safety record that belies the 7 to 10 mph speed variance "efficiency" that now cripples our interstate. However, it is important to point out that Drive Right does not eliminate the problem of speed variance. Some of the more spectacular (and deadly) crashes that do occur on the German freeway system happen when a slower vehicle pulls out in front of a faster one. Drive Right is not infallible. It does help to reduce dramatically the *frequency* of speed variance accidents, but does not remove the problem of speed variance on a high-speed freeway.

Though this debate over the benefits of uniform pace and lower speed variance will not fade from the American highway safety agenda overnight, there is enough evidence in favor of a Drive Right, Except to Pass law to begin moving in that direction. The fact that this rule of the road can help to reduce speed variance and improve traffic flow under certain circumstances should be a cue to traffic engineers, law enforcement and the driving public to support its implementation. In fact it has already begun. In 1997, Texas led the way, passing a law for its freeway system: "Keep Right Except to Pass."

At present, on heavily traveled and congested stretches of U.S. freeway, the flow of traffic within lanes can be dramatically upset if one or more vehicles impedes that flow by driving below the pace speed of that lane. When traffic moves in a proper Drive Right pattern—with slower vehicles to the right and faster ones to the left—the efficiency of the freeway is quite high, or about the best that can be achieved. However, if just one slower vehicle wanders over to the left, but maintains the slower speed of the right lane, the flow of traffic is greatly impeded and efficiency drops. Also, the potential for an accident increases as does the aggravation level of motorists as traffic bunches up behind this Left-Lane Bandit. In this instance, strict adherence to Drive Right would improve traffic flow and speed efficiency while reducing accident risk.

Drive Right can also reduce speed variance under similar conditions by keeping the slowest flow of traffic in the farthest right lane—the lane used for entering or exiting the freeway. For example, if the traffic in the right lane is moving at 60 mph and the left is at 70 mph, a vehicle accelerating on to the freeway merges into the slower flow of traffic. On the other hand, if a slower vehicle slides over to disrupt the 70 mph flow of the left lane and this traffic moves over into the right, there is now a much greater speed variation between this flow of traffic and merging vehicles. Again, Drive Right maximizes the flow of traffic while reducing the speed variance (not to mention the stress levels of those entering or exiting the freeway).

Another important facet of the Drive Right agenda overlooked by the safety community is the benefit to law enforcement and emergency vehicles for high-speed pursuits and lifesaving missions of mercy. For state troopers, paramedics and fire fighters, Drive Right *must* become law to help save lives. Police and other emergency vehicles must cover great distances in short periods of time, and mere minutes can spell the difference between life and death. Drive Right will safely permit law enforcement, fire and ambulance drivers to reach their destinations as quickly as possible. If Drive Right were second nature for motorists on the road, these vehicles could pass by uneventfully.

I have ridden along with more than one state trooper with lights flashing and sirens blaring and have seen people so comatose behind the wheel that they will not move over. Once, I rode along on a hot call, trying to reach an officer whose life was being threatened by a knife-wielding suspect. We never did make it to the crime scene because other motorists would not let us pass. Fortunately, other units were able to get there first and no one was injured, but how many lives have been lost because someone impeded these men and women from their lifesaving duties! This is one legacy of the 55 mph speed limit that we can definitely do without. It is amazing that police, fire and ambulance drivers have not banded together to rid our highways of these inattentive and dangerous Left-Lane Bandits.

Easing America into a better Drive Right attitude over the next few years is a task that is *not* particularly difficult, as long as there is proper law enforcement backing. For most states, a simple change of law on the books and a re-signing of applicable stretches of highway will be sufficient. Ignorance and laziness are the main reasons Drive Right is not more strictly observed. Enforcement is not terribly time-consuming, either. State troopers can begin "Drive Right Sweeps" on their normal patrols, moving down the interstate in the left lane above the flow speed of traffic, getting people used to the

WHY DRIVE RIGHT, EXCEPT TO PASS?

1. Permits high-speed police and emergency vehicles to pass quickly and safely.
2. Allows slow and fast drivers to share the road safely, working together for a better flow of traffic.
3. Makes it easier for vehicles to enter and exit the freeway.
4. Eases stress for all drivers using the road.

How do you indicate to a slower vehicle in the left lane that you would like to pass: Maintain a safe following distance and turn on your left turn signal. If the vehicle doesn't move over—when a break in the traffic comes—flash your headlights on-off quickly (high-low at night).

DON'T BE A LEFT-LANE BANDIT. DRIVE RIGHT!

Possible "tips" brochure for the state patrol to hand out, taking the first step to improve driver's education.

idea of someone coming up from behind to pass. This is one of the few instances when unmarked squad cars can be beneficial to highway safety. Some people will promptly move over when a marked police car is behind them, but will sit in the left lane all day long when other traffic would like to pass. An unmarked squad can effectively seek out this minority of Left-Lane Bandits. And this does not have to begin in a harsh or punitive way. Many drivers do not know what they are doing. Pulling this individual over can be an effective educational tool, using the trooper to inform the driver with a talk and four-point brochure on why the Drive Right law is so important. The officer making this educational stop can record that the driver was pulled over for this problem. If it happens again, this person can be issued a formal warning, and on the third infraction a bona fide ticket can be written with fine and points assessed. This makes the conversion process more palatable, less authoritarian and ultimately more effective. Once Drive Right is explained, most drivers won't block traffic again. Drive Right becomes the first and most important step in upgrading America's attitude about its driving habits, based on the premise that we *all* have to work together to save lives.

Drive Right goes hand-in-glove with a policy of posting real-world speed limits on our highways and freeways. Speed limits well above the normal flow of traffic will help to create—in the short term—a slower-traffic-keep-right condition that lends itself well to the Drive Right conversion. For vehicles moving down the rural freeway in the range of 70 to 75 mph, an 80 mph speed limit will still permit legal passing in the left lane, promoting Drive Right safety and efficiency. Over time, the speed of traffic will begin to creep up near this new posted limit—a trend that we should no longer fight, but accept. Improvements in road and vehicle design, along with drivers' changing attitudes, will cause this slow, but natural, increase in speed. Drive Right will help to align traffic in a way that promotes a uniform flow within lanes. As speed increases, both fast and slow motorists will better understand what is expected of them, establishing the most efficient movement of vehicles possible.

To contrast this type of driving environment with what we have tried in the past, one only has to look back to the days of the 55 mph speed limit on our rural interstates. The fact that some drivers would never exceed the posted limit led to very inefficient and dangerous use of the freeway. Quite often two vehicles would line up side by side, the driver in the left lane unwilling to speed up or slow down just long enough to move over and let faster traffic pass. What soon developed was a 55 mph "caravan" of cars and trucks building up behind these two interstate road racers. Besides causing great stress and aggravation to those mired behind, this condition created a potential road disaster should this glut of traffic become entangled. Drive Right, along with speed limits above the flow of traffic, will definitely prevent such a caravan from building up on the road by spreading vehicles more uniformly and at safer distances over the freeway.

Ultimately, the natural extension of faster driving with proper lane obedience leads in the direction of an *Autobahn*-style freeway with no speed limit at all. Acclimating American motorists to Drive Right opens the door to passenger car, light truck and motorcycle speeds above the 80 mph mark on the rural interstate. Over the first few years of Drive Right conversion, the flow of traffic will redirect itself by speed and lane. On urban freeways, this will reduce the level of stress on the average motorist while permitting higher traffic density with increased flow. From the air, this would look like a steady stream of vehicles moving down each lane, the slowest on the right, fastest to the left. On the rural interstate, the density of traffic is, and will remain, much lower. The pace, or flow, of traffic becomes less perceptible. Here on the open road, great speed can be maintained when drivers comply with the Drive Right law. Higher speed limits will help to set this trend, but moving in the direction of a rural freeway with no speed limit will make clear to all drivers that high speed is the rule and Drive Right makes it happen. On a no-speed-limit freeway, the assumption is that high-speed vehicles *could* be approaching from behind at any moment. And this fast-moving vehicle may not be someone driving an exotic car, either, but a state trooper in hot pursuit or an ambulance trying to reach its destination to save lives.

Once Drive Right and higher speed limits make their mark on the driving public and the driving environment becomes more relaxed and efficient, motorists themselves will voluntarily obey and, more importantly, self-enforce this law. A high degree of compliance with Drive Right is vital to achieving higher speed limits with no reduction in safety. Just a few Left-Lane Bandits can spell the difference between a freeway with increased flow speeds and lower speed variance and one bogged down in a restricted flow of traffic. However, we have an excellent chance for widespread acceptance of this law because the benefits to fast and slow drivers alike are so clearly visible. Further, a majority of drivers on the interstate already do Drive Right. They are somewhat sluggish in responding to it and, may not know exactly why they do it, but as a rule, most motorists will Drive Right if someone faster would like to pass. We must firm up this commitment and educate those who do not understand or who self-righteously believe it is their duty to block traffic. This shift in attitude will do more to reduce the overexaggerated problem of "road rage" than any other measure. As always, sections of interstate that have exits on both the right and left side (this is almost exclusively a problem on the urban interstate) can be signed to permit passing on the right. As time and flow speeds of traffic increase, Drive Right will become easier and more self-enforcing because the flow of through traffic will be higher than that of vehicles entering or exiting the freeway, helping to create a natural Drive Right condition.

It does remain to be seen how Drive Right will affect the relationship of safety and speed on the American freeway. Will it be able to flatten out the U-shaped curve of the accident/speed graph? (See Chapter 1 for a discussion of this curve.) Or will it simply reduce the frequency (numbers) of crashes while the shape of the graph remains the same? A preliminary comparison between the graphs for the *Autobahn* and interstate does show similarities, but comparing the interstate against itself as Drive Right is phased in will be the true test of this rule's effectiveness in raising speed efficiency and saving lives. While only traffic engineers may be interested in this outcome, what should concern every motorist in the coming years is finding common ground with fellow travelers to make the driving experience as enjoyable as possible. Drive Right meets that demand. It is not a cure-all for fast or slow driving with total safety, but just one very important step in the right direction.

Part Two
A Safer Roadside, A Faster Road

Drive Right is more than just the attitude of proper lane obedience. It utilizes the basic layout and design of our highway and freeway network to permit faster driving while maintaining a high degree of safety. This roadside safety program breaks down into a two-tiered system that includes, first, the permanent part of the road itself, including its surrounding right-of-way. Second, an electronic support network that provides safety and emergency information to *all* drivers on the road. These improvements take advantage of technology that is already available, ranging from low-tech signs and markings, to medium-tech communications and sensing equipment. Driving both Fast and Safe does not require any far-out, high-tech gizmos and gadgets, nor must tomorrow's freeway be a technological nightmare of confusing information. To the contrary, the whole idea of road safety must remain simple and easy to understand. It must not intrude on the driving experience, unless there is a clear and present danger. It leaves motorists free to enjoy every turn of the road and crest of a hill, unfettered by information overload. Safety and speed then become a natural by-product of all drivers feeling confident and relaxed behind the wheel, knowing what's up ahead before they get there.

A road is only as safe as its design. Many precautionary measures must be taken in order to minimize, or in some cases eliminate, the accident risk associated with speed. The basic design of a road dictates the limiting factors in a roadside safety program that will effectively reduce crashes directly attributable to higher-speed driving. As stated earlier, the only road, by design, that can be adapted to a no-speed-limit condition is a limited-access freeway. With the two directions of travel separated by a median strip, and no at-grade intersections, the basic layout of the present-day American interstate already moves traffic safely in the range of 70 to 80 mph. However, if speeds are going to increase, as over time they surely will, the limitations of our freeway system will be exceeded. This is not so much a problem of the road surface being unable to handle high speed; it is more a problem of sight distances and the need for more guardrails and energy-absorbing barriers to accommodate vehicles moving above 80 mph.

For starters, a freeway is only as fast and safe as its road surface is smooth and uniform. For high-speed driving, the interstate must be a continuous ribbon of highway, unblemished by potholes, popped expansion joints or buckled pavement. When federal and state gas taxes were raised in the 1980s to help pay for part of this restoration, the interstate system was in need of serious maintenance and repair. Over the last decade, thousands of miles have been brought back up to federal specification, and in the coming years, the entire system will, more or less, achieve a level of consistency to move safely in the direction of higher speeds. Critical for the future is a reevaluation and improvement of construction techniques to ensure that new road surfaces are of the highest standards. Every effort should be made to extend sight distances over hills and around curves, taking advantage of greater superelevation (banking). Mating these improvements to a stout road base with ripple-

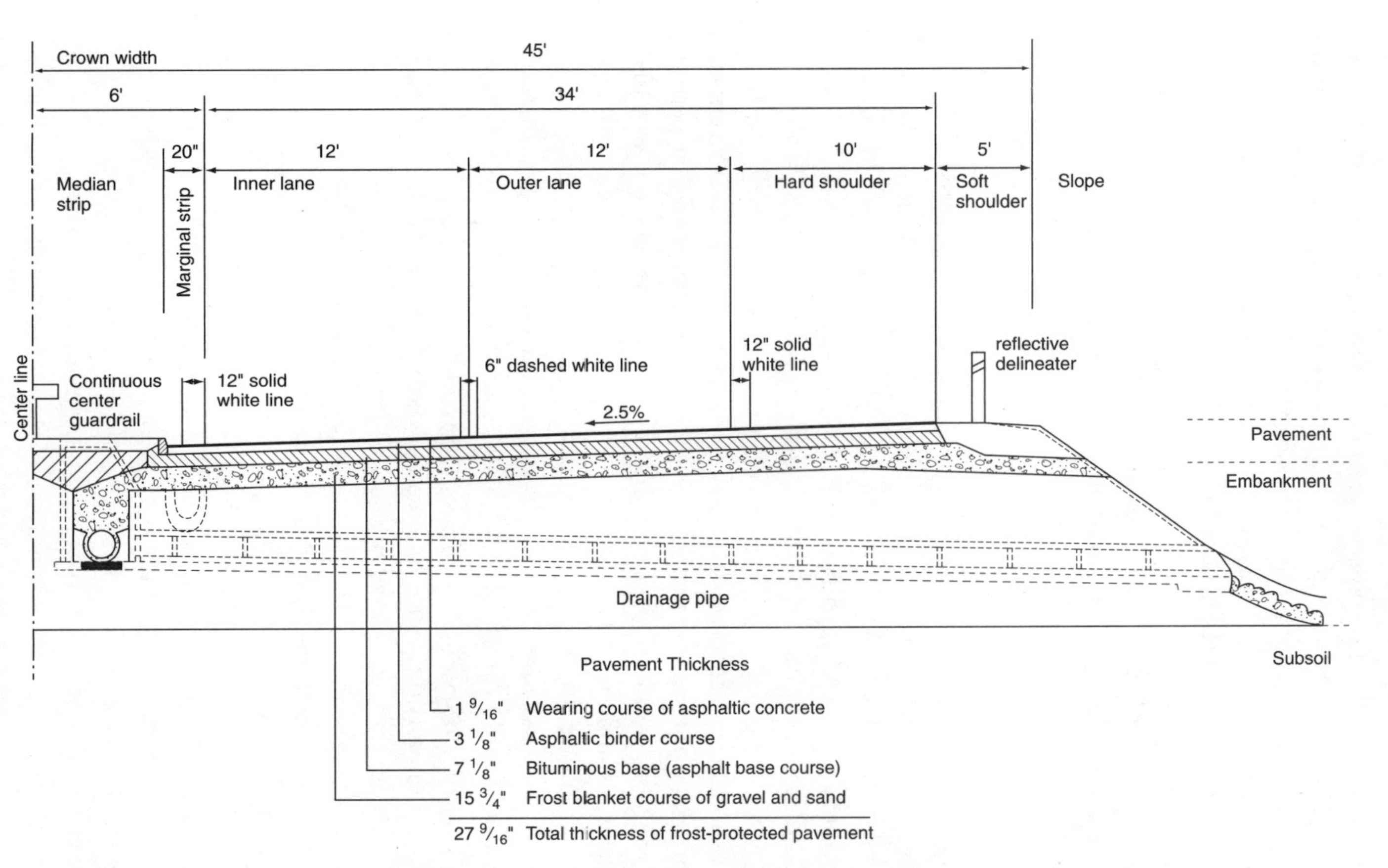

Cross section of a typical segment of German Autobahn.

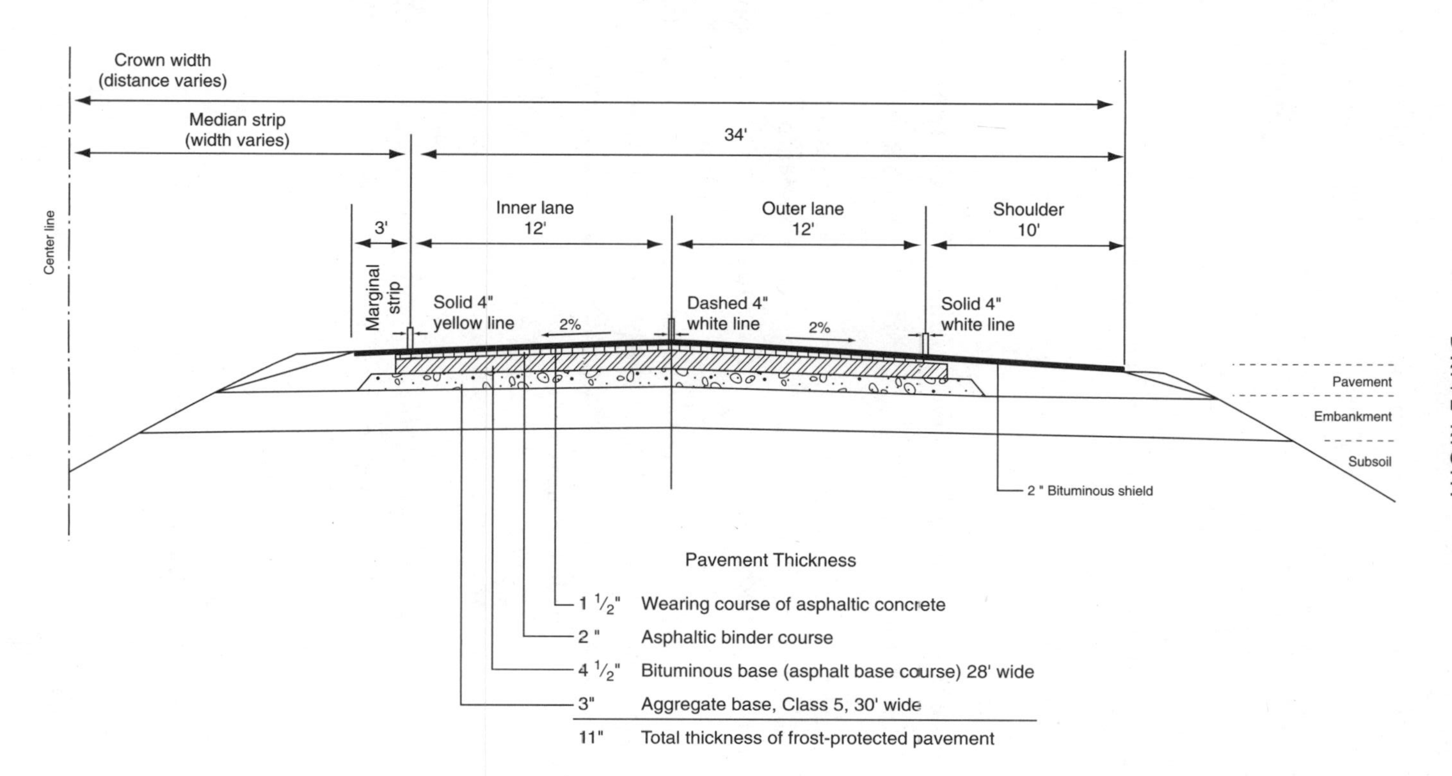

Cross section of a typical segment of American interstate.

free pavement maximizes safety and speed efficiency, and minimizes wear and tear on tires, brakes and suspensions. Building highways to last, with proper original construction and timely maintenance, extends the period of total overhaul and replacement, which saves tax money in the long run.

The *Autobahn* roadbed is more than two feet thick with an expected service life of forty years. Across its entire width from the inner driving lane to the edge of the shoulder, the foundation is one solid, frost-resistant structure that prevents cracking and warping. It receives regular maintenance to keep the upper driving surface smooth and free of potholes. At $12 million per mile, the initial construction cost is fairly high, but the roadbed lasts twice as long as the typical stretch of interstate costing $10 million per mile. The German freeway also moves more vehicles at higher speeds and with less of a disruption to the traffic flow that is caused by road repair. In the long run, this better road design is cheaper. However, America's road-building techniques are steadily improving. Recent developments in "superpave" asphalt and better sensing equipment to lay down the road surface are raising the level of smoothness and service life for new highways. Unfortunately, we still have a long way to go to match the outstanding engineering of German and European roads in general.

Once this flawless road surface is established, it must be painted and signed to provide the most straightforward, easy-to-understand driving environment possible. On the *Autobahn*, there is never any doubt which lane you are in or which direction the freeway is headed. Even at night in a driving rain or snow, the outline and shape of the German superhighway are apparent. This is one area where the American interstate can be immediately improved by repainting the road surface to indicate more clearly the location of the driving lanes, entrance and exit ramps. It is appalling in some cases to see how sloppily and shoddily the lines of the interstate are laid down. These are sometimes so poorly placed on the freeway, it looks like a drunken highway worker has walked down the roadway with a bucket and brush. I once saw a line painted right over an abandoned tire on the shoulder of I-94 near the Mississippi River in my hometown of Minneapolis. Further, the lines on the U.S. freeway system are, in general, much too thin and stringy to be of benefit to motorists, especially as speed picks up. The lines on the *Autobahn* are always a clean, bright paint. White markers at exits and arrows indicating turns are imbedded in the pavement and have an extremely long service life. These can, in turn, be painted over, providing a ready-made template when touchup is required. The lines on the shoulders of the *Autobahn* are a foot wide to orient drivers clearly to the direction of the road itself. A dashed version of this bold line highlights where the entrance and exit ramps begin and end, keeping them separate and distinct from the driving lanes. There is no mistaking the shoulder of the German freeway, either. It is a clearly defined place where stranded or emergency vehicles can pull over. These lines are precisely and uniformly spaced on the road surface. More importantly, when repainted, great care is taken to place them in the exact same spot as before to avoid creating a "ghost image" of faded or oversprayed paint that could confuse motorists. The lines don't waver or flutter in a jerky fashion that can become fatiguing or dangerous on a long trip. Such attention to detail spells greater speed in more comfort, especially at night when these heavy white lines reflect more light back over a greater distance. Emulating these paint techniques on the American interstate are a quick and easy way to provide cost-effective safety at higher speeds.

Even bold lines tend to lose much of their visibility by night, and can be totally obscured in rain, slush or snow. For driving at night, or in bad weather, reflective delineators on the right shoulder further outline the edge and direction of the *Autobahn*. Mounted on "shear-away" posts at uniform intervals, these reflectors return the faint glow from far-off headlights. For fair-weather, nighttime driving, this dramatically improves visibility at high speed. As the painted lines on the road surface vanish into obscurity, the delineators on the right shoulder extend a driver's line of sight much further, indicating the freeway's direction with an ever approaching stream of reflective, white points.

Expanding on this premise for either the *Autobahn* or interstate, the perimeter of bridges, exit and entrance ramps could be outlined with reflective material to upgrade visibility. Many heavy trucks and, especially, school busses already use this highly reflective tape to highlight their exteriors at night. Placing red reflectors on the

Autobahn versus interstate. Notice the clear delineation of the lanes, including entrance and exit ramps, the outline of the road itself highlighted with reflective posts. Even at night in inclement weather with snow-covered pavement, the shape and direction of the roadway are apparent. A heavy, double guardrail separates the cross traffic. Foliage growing through this guardrail acts as a natural antiglare fence against oncoming headlights. All these preventive measures add up to increased safety at high speed. . . Photo by the author

However, all is not lost on the American interstate, either. The right-of-way is much larger with a wider median strip. Trees and shrubs are farther away from the road's edge. The basic design of our freeway system is more than ready for an Autobahn-style roadside safety program to aid faster driving. American Autobahn Society Archive

backside of bridge abutments and on roadside delineators might indicate to the handful of confused or drunken drivers that they are traveling the *wrong* way down the freeway. The worst type of collision on any divided highway is when someone drives in the wrong direction. In more than twenty-five years on the road, I have seen this happen only twice—once by an older driver who was confused, and the other time by a drunk who caused a gruesome crash. It is a rare occurrence, the possibility of which must be reduced even further because of the more deadly consequences at higher speed.

Presently, the most serious roadside safety deficiency on the American interstate is the dramatic lack of guardrails. This is almost exclusively a problem on the median strip which separates the cross traffic. Budget cuts and the desires of state troopers to keep this median open to nail "speeders" perpetuate this hazard. Even at today's modest speeds, when an errant vehicle jumps the median and strikes another one head-on, the collision is horrendous. For high-speed driving on the interstate, there must be a continuous guardrail on each side of the median strip.

For a better understanding of the importance of this type of roadside safety, we can learn from the improvements in racetrack design and apply them to public roads whenever possible. A solid, continuous barrier is just one such innovation. When driving at high speed, it is critical not to run off the road, or track, under any circumstances. Any time you can continue in the same direction and not strike a solid, immovable object, you are better off in terms of safety. The one object you *can* strike is a solid, continuous guardrail, running parallel down the edge of the road. Here, a glancing blow lets you scrub off speed and slow down progressively after a collision. Even hitting other vehicles that are traveling in the same direction is safer than leaving the road to strike a tree, a bridge abutment, or a vehicle in the opposing lane of traffic. Shielding both sides of this median strip effectively puts an end to the possibility of vehicles jumping over into the opposite direction of travel. This is one advantage the interstate, with its larger right-of-way, has over the *Autobahn*. In high-speed crashes, vehicles have been known to vault over the heavy, double guardrail on the German freeway. A very narrow median strip permits this. On the interstate, with guardrails on each side of our wider, center strip, the possibility of vaulting over the rail on one side still exists, but the width of the median and the use of another, separate guardrail on the other side would make a head-on collision all but impossible, even at very high speeds.

For guardrails placed on the right shoulder of the freeway, the point where the guardrail ends is of particular concern. It must be protected with an energy-absorbing barrier to prevent vaulting over, and to keep the end itself from crashing through the passenger compartment of a vehicle. Designs currently used on the interstate do not provide adequate protection from these problems, even at today's speeds. The end of the guardrail must be covered with an extensive sandwich of foam, plastic and steel that turns it into a progressively cushioned unit, preventing serious injuries or death—even at speeds approaching 100 mph. The increased cost of such a deformable end will probably make extending the guardrail to the next section cheaper by comparison. Such a crash-absorbing endpiece will then be used only sparingly. Here again, the racetrack might provide an extremely affordable solution to emulate: old tires stacked together at the guardrail's end, painted in bright colors and topped with plastic to keep out rain, snow and animals. Whatever method is used, at $6 to $10 per foot, a guardrail is one of the more cost-effective solutions to shield obstacles and prevent vehicles from either leaving the freeway or jumping over into the opposing traffic.

There is another vital piece of safety equipment that is used far too infrequently on America's freeways: A series of green cones mounted on top of the median strip guardrail to eliminate the glare from oncoming headlights. This "antiglare" fence then permits motorists to continue using their high beams to increase speed without adversely affecting safety. This becomes a double-win scenario as visibility improves along with line of sight. Speed does not have to be sacrificed because of restricted vision. For the standard stretch of interstate, only one side of the median guardrail need be equipped with these cones as long as care is taken to switch them from side to side, depending on which direction the freeway turns. This prevents the possibility of the antiglare fence itself restricting sight distances down the road. For curves to the left (in your direction of travel), it is important that the cones switch to the opposite side so as not to

"Texas twist" guardrail. The end of the rail is anchored directly to the ground, preventing it from crashing through the windshield into a vehicle's passenger compartment, but the "twist" does not stop a vehicle from launching itself into the air on this ready-made guardrail ramp. American Autobahn Society Archive

"Energy-absorbing" guardrail. This updated design solves the ramping problem but will require further energy-absorbing modifications to deal with very high-speed crashes into the end. American Autobahn Society Archive

restrict line of sight. You can still see over the top of the guardrail in a left-hand turn and have time to react to any potential problems further down the freeway. For motorists traveling in the opposite direction, the road almost always follows the reverse curve, so the fence mounted on this side does not interfere with sight. For extremely tight turns, which are almost nonexistent on the interstate, the cones can be spaced closer together or filled in to prevent distracting glare from spilling over to the other side. Besides being safer at night, this antiglare fence can help to prevent second or third "gawker" accidents, day or night, because no one can see the aftermath of a collision in the oncoming lanes. With all these innovations, each mile of interstate takes on a state-of-the-art appearance to reduce accident risk.

Of course, all of this roadside safety requires first-rate maintenance to keep the freeway in tip-top condition, especially after a crash has occurred. Damaged guardrails, reflectors and antiglare fence could cause another accident, instead of preventing or minimizing one. Repairs must be made promptly, returning the road to its original uniformity as soon as possible. A high-speed freeway

Antiglare fence. These green cones on top the guardrail prevent annoying and dangerous headlight glare from shining over into the opposing traffic at night, permitting higher speeds with greater safety, and allowing motorists to continue using their high beams. American Autobahn Society Archive

should appear as a clean, bright, endless flow of road, keeping your interest and enjoyment level high as you cruise at a comfortable speed, not having to guess where you and the road are going. Any disruption of the road surface, paint, guardrails or signs could call that into question. A less than perfect roadway breeds a lack of confidence in those behind the wheel. Removing oil, coolant and debris—the remnants of an accident or mishap—and conducting regular maintenance and repair instill a confidence which is not false or misleading.

As time goes by, modest technological advances will only enhance this feeling of security. Reflective delineators that clean themselves periodically and pop up after being run over are just as feasible as a "smart" guardrail whose end can be reset back to original shape after it has been struck. Such devices would be cheaper to install and operate than constant replacement with new parts. These are just a few of the innovations possible in a cost-effective roadside safety program that improves faster highway transportation.

There is another problem with the roadside on the interstate that stands in the way of very high-speed driving. The solution to this problem lies not in a physical improvement to the freeway but in an electronic one. The original design parameters of the interstate set down a working speed range for vehicles in the neighborhood of 70 to 80 mph. It is important to remember that this range was determined back in the 1950s, using the standard automobile of the day. With the dramatic improvements in tire, suspension and braking technology, this range of speed can be upgraded to well above 100 mph for most of the interstate system. On some segments of the freeway, however, a line-of-sight problem still remains. When a vehicle travels ever faster down a road there becomes less and less reaction time to see over a hill or around a curve. At very high speeds, motorists could overdrive their ability to slow down. While this occurrence is rare on the interstate, the problem does exist and precautionary measures need to be taken. The simple solution of a reduced-speed sign is not sufficient. The temptation to continue on through this curve or over this hill at high speed is too great. The solution to this danger lies in technology that has already been developed as part of the federal government's Intelligent Transportation Systems (ITS).

ITS has been planned for several different areas of highway safety to improve traffic flow, and reduce congestion and pollution on most urban stretches of freeway. It goes by several different names across the country, with programs already on the road in Dallas (DART), Chicago (ADVANCE), and in Minnesota (Guidestar). Over the next decade, the federal government will spend more than a billion dollars to test and implement these programs nationwide. It is estimated that over the next two decades, $200 billion in private and public-sector money could be spent to further develop the concept of intelligent highways. While many of the proposed programs will have a positive effect in reducing accidents, gridlock and in cleaning up the air, others are of dubious value and a couple are outright dangerous. Further, any such program is a complete waste of time if we don't permit faster travel on the open road. If we begin to deal with the risks and benefits of speed, ITS opens the door to innovations that will make high-speed driving with improved safety a reality in just a few years.

The line-of-sight question is the first speed-related problem that ITS can tackle on two fronts, beginning with a one-way communications network that will quickly and accurately inform motorists as to possible dangers ahead. The Advanced Traveler Information System, or ATIS, has its roots in a similar system now used in aviation. Using the same initials, the Automatic Terminal Information Service provides up-to-date weather and landing information for larger airports on a recorded loop assigned to a specific radio frequency. Before taking off, or when approaching an airport, a pilot can be properly informed as to weather, traffic patterns and runways in use. Aware of what is happening around the airplane, the pilot can adjust the flight path to existing conditions. Speed is not a part of the equation. On the freeway, the same can be true. If you know what is up ahead—construction, a fender-bender, or gridlocked traffic—you can plan accordingly by slowing down or taking an alternate route before trouble turns into an accident. Broadcast over FM radio on the low-end of the spectrum, this prerecorded information is updated as needed. As in aviation, call letters of the alphabet can be used, starting with information *A*lpha. This continuous loop of weather, road conditions and alternative route information is readily available to plan a trip or update one on the way. At present, many metropolitan areas already use a basic version of this system on commercial radio. As time passes, this will be improved to a more detailed and user-friendly separate radio station to aid driving. Depending on the region of the country, large cities could have several frequencies to monitor different routes. Large states like Wyoming and Montana with small populations and few freeways might need only one. The success of the system will come from its ease of use, developing the technology so *every* motorist can access it. Acceptance will depend on not having the system intrude on the enjoyment of driving, unless a true emergency exists.

One of the present deficiencies of ITS is that it reaches only a handful of vehicles equipped with high-tech gizmos and gadgets to test and prove the technology. Dashboards crammed with expensive TV screens, satellite navigation systems and two-way radio communication are, at best, nifty, futuristic novelties that do nothing to

The rudimentary beginnings of the interstate freeway A.T.I.S. system are already in place. American Autobahn Society Archive

improve congestion and speed efficiency for *all* drivers in a cost-effective manner. These devices can also set a dangerous precedent. The last thing you want in a car is a dash-mounted television. Imagine driving down the freeway at 80 mph, one hand on the wheel, the other pushing buttons on a TV, searching for directions. If this trend continues, watch for an entirely new breed of auto accident in the coming years where motorists are killed and injured because their heads are buried in an ITS TV screen! This type of technology should be outlawed for all vehicles, save those of law enforcement officers, who need a computer for the performance of their duties. To be of benefit, information must be available to all motorists on the road, focusing their attention on the road itself, not inside the car. This limits technology to road signs or audio information so drivers can listen while eyes are still on the road. This is absolutely vital to safety at any speed. Further, view-screens in cars will always be much more expensive than an advanced FM radio receiver.

The Automated Highway System is another boondoggle that would spend millions now, perhaps *billions* in the near future, to test a fleet of cars that can drive itself. While some of this technology has worked its way into many automobiles (antilock brakes, traction control), the immense expense is a waste of public and private-sector money, which would be better spent putting the other more valuable, and presently feasible, ITS technologies on the road years ahead of schedule for all motorists to use.

The ITS ATIS program opens the door to many diverse possibilities to inform motorists as they drive down the road faster. But in the short term, more programmable signs and flashing lights can alert drivers to a change of condition. As the FM radio technology is further deployed, low-power antennas would be placed at each entrance/exit ramp to give several miles of coverage, both on the freeway and the roads approaching the interstate. The normal radio in a vehicle could then be tuned in to the proper frequency for the latest recorded message on road and weather conditions. In just a few years, this could be refined to have the radio turn on automatically a certain distance from the freeway, or when a new "loop" of information is fed into the ATIS system. State troopers might tap directly into the network for a live emergency broadcast within a two- to ten-mile perimeter the instant a problem occurs. No longer would motorists blunder into a dangerous situation. With the preponderance of cellular car telephones, a 911-like number could also be put into service so information about trouble could be relayed directly from a motorist to the nearest state trooper, and be added quickly to the ATIS system. The time could even come when one side of the freeway is informed of a potential problem while the opposing traffic speeds on unimpeded. Such a safety program, using one-way radio communication enhanced by cellular phones, would create a cost-effective way to keep everyone on the freeway informed, helping to avoid and minimize accidents while speeds pick up dramatically.

In the broad sense, the ATIS system gives timely and up-to-date reports so motorists have a feel for what is happening on the freeway ahead. In the specific areas where hills and curves cause a line-of-sight problem, another low-cost, medium-tech device can facilitate high-speed driving with enhanced safety. If you knew before rounding a corner or cresting a hill that a problem existed on the other side, you could plan accordingly, either by slowing down or maintaining your speed. A line-of-sight warning indicator could provide such information on those rare hills and curves that restrict vision at high speed on the interstate. This device could be as expensive and complex as sensors buried in the road surface (these magnetic loops are already used to record speed and trigger traffic signals), or a more modest radar unit recording the speed of vehicles passing by. This warning indicator senses the speed of vehicles through these problem hills or curves and, with the aid of a small microprocessor, monitors the speed variance. If one vehicle is closing too fast from behind in relation to slower ones up ahead, a series of flashing yellow lights or a programmable sign could inform the speeding motorist to slow down. Under normal driving conditions, even at very high speed, these lights and signs would not flash a warning—indicating to this fast driver that all is clear ahead. A person can drive with a real sense of safety and security, keeping up a much higher average speed over a long trip instead of braking and slowing down just to make sure (or foolishly taking the risk and hoping for the best). If the unit malfunctions, a backup power supply or continuous flashing yellow lights would prevent a possible catastrophe, making the device all but foolproof. Such a line-of-sight, speed variance monitor would aid fast and slow drivers, raising confidence levels in both.

An offshoot of this technology, and associated with the ITS program, is the Variable Speed Limit System. Positioned on the freeway after every entrance ramp, each speed-limit station uses a magnetic-loop sensor in the road surface. A computer bases calculations on the average flow speed of traffic to display a real-world speed limit for the given conditions. On the urban interstate, this can help move vehicles far more effectively than a posted limit on a static road sign because the variable limit "floats" with traffic and weather conditions. When the road is wide open in fair weather, there might not be any speed limit at all, and the freeway can be used to its maximum capacity. As traffic builds, the limit adjusts accordingly, again maximizing speed and capacity in relation to traffic flow. This has great potential for easing congestion and gridlock because it more efficiently speeds vehicles along, using both minimum and maximum speed limits to create a uniform flow. This might also help shorten the rush hour of peak traffic because the vehicles just before and just after could be moving at a much higher speed than is presently allowed on our urban freeways. Yet, as traffic builds towards the peak hours, the limit automatically would lower to prevent an erratic flow of high- and low-speed traffic from bunching up (an undesirable condition that can cause serious accidents). A similar system used on heavily traveled segments of the German *Autobahn* has reduced accidents, in some cases, by 30 percent.

The Variable Speed Limit System with the moderate price of $40,000 per station has the potential for permitting higher speeds with lower accident risk. It is also important to note that much of the urban interstate system and certain rural segments could benefit from a variable speed limit to maximize the movement of traffic in relation to speed and safety. However, there are vast tracts of rural freeway west of the Mississippi River that require no speed limit signs at all, just the proper precautionary measures to make high-speed driving safer. This is partly due to the lower number of vehicles using these stretches of interstate, but there are also varying

Variable Speed Limit Sign on I-40 in Albuquerque, NM, 1988. This computer-controlled system uses magnetic sensors imbedded in the pavement to base a speed limit on the average speed of traffic, plus or minus so many miles per hour, taking into account congestion, traffic flow and weather. In the photos above, the speed-limit sign has been turned away from traffic to simulate what the real-world speed limit would have been under various conditions back in 1988. With the national 55 mph speed limit finally gone, this promising innovation can now be used to speed people and commerce with enhanced safety. Photos by Nevin Harwick

regional differences toward speed control that must be considered. In the East, in general, citizens are more comfortable with a maximum speed limit to govern traffic. In the West and South, drivers and officials have historically taken a more relaxed attitude. This has sometimes led to a fair amount of hypocrisy with all parties involved—either trying to crack down on speed enforcement or to look the other way. But these attitudes must be factored in when dealing with America's overall perception of the problem. To that end, the Variable Speed Limit System might be the best compromise for safety and speed in many parts of the country.

The physical and electronic reevaluation of our freeway system demands that faster travel be a part of the equation. The immense expense of upgrading its shortcomings can only be justified when safe, high-speed operation of all vehicles is taken into consideration. Safety philosophy must not be based on trying to slow motorists down simply because there *might* be danger ahead. The danger must be removed or clearly identified to minimize accident risk—without sacrificing speed. With that philosophy as the goal, tomorrow's roadside safety improvements on America's highways become more desirable and cost-effective. Speed is then just another vital interest in the process of building better roads and maximizing transportation efficiency.

CHAPTER 7

An A.D.E.P.T. Driver: Safety and Speed through Education

Driving in America is neither a privilege, a necessity nor a skill. It is a birthright. That is the attitude governing our perception of the training citizens should have to operate a motor vehicle. This heritage has its foundation in the horse-and-buggy days from which the automobile developed. No formal training was required to hop on a buckboard and ride off with a team of horses. Driving a car evolved out of that tradition, compounded by the fact that America was desperate to get out of the knee-deep mud and horse manure covering its streets. The booming prosperity that came with the automobile demanded easy access to driving. Over the years, the licensing of motorists moved the country slowly towards recognition of the need for some kind of basic driver-skills program. Starting out as word of mouth and as technique handed down from parents to children, by the 1950s, driver training had grown into a more formal program taught privately as well as publicly through high schools across the country. This has been mediocre at best.

When my mother went to city hall just after World War II to pick up her driver's license, that was all there was to it. First, however, she had to pass my grandfather's personal driving school of city traffic, country highways and changing a tire all alone! Almost thirty years later, my "official" training program consisted of endless dos and don'ts, sterile and boring lectures with the usual bloody crash films and a modicum of behind-the-wheel training to show "proper" operation of a motor vehicle. Fortunately, my father also instructed me in the ways of real-world driving. To this day, the same driver's training program is in place for sixteen- and seventeen-year-olds. Once you reach the age of eighteen, you don't even have to take any kind of training in most states. Passing a written test and basic skills exam—several turns to the right and left, stopping and parking—are all that is required. Training for trucks, buses and motorcycles varies from fair to nonexistent. In a word, driver's education in America is a joke. The only positive aspect of this training is that the average American is *not* an idiot but takes this basic education and develops it over time into relatively good driving ability.

The need to improve and update driver education for all motorists has been met with talk and little else by the driving public, politicians and, unfortunately, the highway safety community itself. The expense, lack of political will, and the view by many auto safety advocates that driver's training programs are of dubious value have helped to create stagnation in the field of improved education. Safety "experts" will point to the lack of statistical evidence that better training produces a safer driver. Within the confines of the Slow-is-Always-Safer philosophy, such an assessment couldn't be more true—viewing a "good" driver as anyone who drives 55 mph or less. The present safety movement in America does understand one aspect of improved driver education that it does not want to acknowledge. A well-trained driver is a *fast* driver on the open road, and this does not fit in well with a philosophy that is constantly trying to slow people down. However, in the world of Fast and Safe, improved driver education is not a hindrance to safety but a virtue and a necessity.

The problems of expense and implementation do still remain, however. Political will and public consensus, while difficult to achieve, will certainly be more receptive to the idea of putting a better driver on the road if we begin to drive down that road ever faster. How to do it will be the most difficult task of the entire Fast-and-Safe agenda. We will never be able to completely retrain and improve every driver on the road. While it would be nice if everyone could go through a complete, comprehensive driving course so we would know, once and for all, that every driver on the road is properly trained, that is beyond the scope of feasibility. Yet, every driver can participate in improving courtesy, adherence to the rules of the road, and safety education, becoming over time an ADEPT driver.

Advanced Driver Education Program and Training (ADEPT) would be a system of improvements that starts out small, looking for modest results in upgrading the average driver's knowledge and ability behind the wheel. It would also develop a long-term strategy, dealing with the requirements necessary to put highly trained, new drivers on the road who will be operating their vehicles at ever higher speeds over the coming years.

The present-day emphasis on improved driver's training does not deal with the majority of motorists on the road. The average driver is not included in any training program while certain segments of the motoring public such as truckers, bus drivers and motorcyclists are singled out for "improvements" to their training. Since these motorists represent a small percentage of total drivers, the results of these improvements are meager at best. ADEPT would continue to deal with this problem in similar fashion, but expand its scope to all drivers.

Our rush to permit virtually anyone to drive has helped to create a transportation network that is now regulated at the lowest common denominator. Anyone can drive, but only at modest speeds with marginal safety. More importantly, this mediocre driving environment forces good drivers and individuals who use the system the most (truckers, over-the-road salespeople) to sink to this inferior level, instead of requiring the occasional user to step up to a higher skill level and better understanding of the rules of the road. Here is where we can begin, and driving faster gets us there quicker.

Three areas of improvement will move America in the direction of advanced driver education. The first is the Drive Right, Except to Pass program previously explained. The second moves seatbelt use near the 100 percent mark over the next five to ten years. And the third establishes a national program of brief television segments to enhance driving skills and knowledge of the rules of the road.

Within the context of driving Fast and Safe, the need for seatbelt use is an absolute necessity. Every man, woman, child, dog and cat must be buckled up to save lives. Sudden braking or an emergency evasive maneuver means unbelted drivers and passengers are going to bounce around, inevitably leading to injuries and loss of control. In a crash, not being strapped in guarantees serious injury or death. Unfortunately, being required by law to wear a seatbelt is clearly an infringement of personal freedom. However, there is no other way to reduce dramatically the number of people killed and injured on the road unless everyone is buckled up. Forty-nine states already have some kind of seatbelt law (only New Hampshire does not). Unfortunately, most of these seatbelt laws are weak and without any real incentive to buckle up. Still, with this patchwork of wishy-washy laws, 60 percent of Americans already use seatbelts, and that is a dramatic improvement over the 10 to 15 percent who wore them twenty-five years ago. Education about their effectiveness has helped to raise compliance, but a law, fine, and enforcement will ultimately be needed to raise seatbelt use from 60 to 99 percent.

Again, looking to Germany as an example, the same state of affairs existed on its roadways in the 1970s. Seatbelt use hovered around the 60 percent mark with a "recommended" seatbelt law—no fine or enforcement. In August 1984, that changed dramatically as a new law with both fine and enforcement came into being. Compliance rose quickly, jumping from 59 percent in spring of that year to 91 percent by fall for the driver and front-seat passengers. Between 1984 and '85, the number of people killed on German roadways dropped from 10,199 to 8,400—1,799 fewer people. In the following years, seatbelt compliance peaked near 96 to 98 percent for the front seat. After July 1986, its seatbelt-use laws were extended to rear-seat passengers as well, with compliance fluttering around 70 percent.

Such a program must be extended to America's roadways as well. Any state interested in high-speed driving with improved safety should implement a comprehensive seatbelt enforcement program. Virtually all state police agencies presently stand behind the use and benefits of seatbelts. Trading speeding tickets for seatbelt enforcement will be a far easier job with much higher safety dividends. Given that a majority presently do buckle up, and that the number will voluntarily rise with an increase in speed, the remaining few can be stopped and given a ticket. Primary enforcement of this law with points assessed to a driver's record is the key. Only when a trooper or police officer can stop a person solely for not buckling up will the compliance levels for seatbelt use make the necessary jump. Points on a driver's record reinforce the process with an increase in insurance premiums and ultimately the loss of license. A high percentage of seatbelt use is critical to reduce dramatically the number of people killed and injured on the road. Only when 99 percent of drivers and passengers, in fact, buckle up, will the benefits become clearly apparent.

This kind of compliance will take time, however. But tying speed increases to seatbelt use directly will move compliance that much quicker. The psychological benefit of getting something back in exchange for giving something up should help. This is not just one more law shoved down your throat, but a *quid-pro-quo* arrangement: safety combined with reasonable and prudent speed limits. Given the fact that almost everyone "speeds" and a majority already wears seatbelts, the trade-off sets up a more compatible relationship between the average motorist and law enforcement, helping to build camaraderie and common cause.

In this vein, wearing a helmet while riding a motorcycle is *not* necessary for safe operation. It is not comparable to the use of seatbelts. Even if you don't have a crash, seatbelt use is necessary to maintain control of a vehicle to *avoid* accidents. Being strapped in prevents the driver and passengers from sliding around. This is not true with a helmet. Helmet-use laws do not have the same pressing urgency as seatbelt use from coast-to-coast. Improved motorcycle safety at high speed will *only* come through better education of riders *and* drivers. Even at today's highway speeds, a biker (wearing a helmet or not) colliding with anything means all-but-certain death or, at best, a lifetime in a wheelchair from a spinal injury—something a helmet can't protect against. However, twenty-six states already have helmet laws covering all riders and that number might increase when some states raise their speed limits. But the best way to reduce the number of motorcyclists killed and injured in the future is to strive for a crash not happening in the first place—a challenge for riders and safety advocates alike to upgrade the skill level and judgment of those on two wheels. Not wearing a helmet is a very precious right/privilege to many motorcyclists. Establishing an advanced rider training program would be an excellent compromise to preserve this freedom.

This movement towards better driver education must begin to take advantage of the positive role the media can play in reaching every motorist. America needs a *Seventh Sense* television program like the one in Germany. These two- to three-minute safety spots are some of the highest-rated programming on German TV. They are highly successful because they deal with tough, uncomfortable driving situations and offer tips that can truly save your life. Such driving tips will be a dramatic departure for the safety movement in America which has been on a broken-record quest to remind drivers of all the things they shouldn't do behind the wheel. No one is listening. The success of such a program in America will depend on its content and the time it is aired. Placing the show before, during, or after a highly rated show like *60 Minutes* would guarantee a huge national audience, so that millions of Americans would be informed on topics ranging from crash avoidance to cell-phone safety. Keeping it short and sweet and dealing with real-world problems on the road will keep people's interest. Another sure-fire draw for a large audience would be celebrity hosts such as Paul Newman, Tom Cruise or Jeff Gordon—taking their fame and knowledge of race-car driving from the TV screen to public roads. Over time, this spot will slowly raise the level of skill for the majority of motorists on the road. If it sinks to finger-pointing, "slow down" and "don't do this or that," this informational segment will be a complete failure. No one will watch. On either side of this brief program, there would be an excellent opportunity for automotive manufacturers to advertise and thus pay for this spot, showing off their latest performance and passive safety features to entice buyers into the showroom. The *Seventh Sense* would be well suited for those taking driver's-ed privately or in high schools across the land. If it is aggressive enough in its approach and scope, many Americans would begin to discuss its content, whether at home or work. These dialogues between drivers would spread the word to those who did not see the segment. This type of television program is a long-term plan to inform and educate motorists on how they should safely negotiate their way through potential problems on the road. It will not provide immediate results. Raising the education level for all drivers is an enormous task. But through small, steady, incremental improvements, positive results should begin to appear in five years. A *Seventh Sense* TV spot would play an active role in making this happen.

Three segments of the driving population require further scrutiny to improve overall safety on the road: young drivers, older motorists and interstate truckers. Today, the trend spreading across the country is to restrict driving for young provisional motorists, ages sixteen and seventeen. This graduated licensing program, in general, starts with adult supervision and a six-month learner's permit. Following that training period, there is another six months with no nighttime driving unless accompanied by an adult. Finally, after a yearlong intermediate period under varying levels of supervision and with no traffic violations, the driver earns a full license at age eighteen. Two states, Washington and Michigan, are trying to expand on this basic, three-tiered program by requiring more comprehensive classroom and behind-the-wheel training. With all of these licensing programs, the goal is to reduce the death rate for these young drivers, which is twice as high as the rest of the motoring population. Unfortunately (and with the hope of being proven wrong), I doubt that these restrictions will have much effect on lowering the number of young people killed on the road. Immaturity and poor judgment are the main reasons sixteen- and seventeen-year-olds die behind the wheel. No amount of training can remove the false sense of invincibility many teenagers have. However, with the insurance industry standing behind graduated licensing, its continued implementation throughout America is all but guaranteed. The test over the next decade will be knowing if it is truly working, or if it is just a convenient way to "look" safer so insurance companies can still charge the traditionally high premiums to the parents of these teenagers and pocket the profit. The Fast-and-Safe solution to this problem—and it *is* a solution—is the same as it is in Europe: waiting until age eighteen to drive. This is a bitter pill for teenagers to swallow, and it will also be exceptionally difficult to implement state by state. However, it is the only way to eliminate the deaths caused by this risky subset of young drivers. Sadly, thousands of teenagers will probably die before the country turns away from graduated licensing towards granting a provisional license at age eighteen. But, if we truly love our kids, we'll recognize that having them wait gives them a much better chance to enjoy the open road well into old age.

At that other end of the age spectrum, older Americans will need to participate in the Fast-and-Safe agenda as well. This does not mean they will be banished from the freeway in the name of safety. To the contrary, many seniors are already outstanding drivers. When I lived in Germany while researching this book, the couple I rented a room from were in their late sixties and early seventies. They would quite often take the *Autobahn* on weekend trips around Bavaria. She would drive in the range of 75 to 85 mph when conditions permitted, he slightly slower. Their driving speeds were based more on personal comfort than on trying to make time. Here in America, the same is often true. A friend of mine tells of his annual winter trip from Minneapolis, Minnesota to Lake Havasu City, Arizona, beaming with pride that "on the freeway, I drive my age." He turned 84 in 1999, and so far has eluded the highway patrol.

However, some older drivers do create problems, and a system to monitor them needs to be developed. The goal of this program is to give older drivers as much mobility as possible, using restrictions only sparingly to allow them the independence to remain active. Already, some states offer a driving class to show seniors how to deal better with the hectic pace of today's traffic. Tips include driving at off-peak hours and taking only right turns, if possible, to reach their destinations. Such advice helps them to cope with the stress arising from a more aggressive driving environment. The reward for taking this class goes beyond the simple knowledge of skills to a reduction (usually 10 percent) in insurance premiums. As the pace picks up on our highways and freeways over the next decade, programs like this need to be expanded from state to state, with the agenda refined to include such things as Drive Right and ways to relax and be comfortable with speed on the freeway. Beyond this more informational class, there also needs to be various limitations on some older motorists' use of the car. This is not a campaign to hunt down old drivers gone bad, but a family, community and law enforcement-based program that works in stages to provide maximum mobility for as long as possible. For example, a problem as simple as poor night vision might be addressed by restricting driving to daylight hours. Many states already have these restrictions but use them far too little. Restrictions such as no freeway driving or intrastate driving will only become more common as we begin to deal more openly with the risks of higher speeds. The decline of driving ability with advanced age or failing health should be handled in progressive stages, giving older motorists plenty of warning that there is a problem without immediately taking away the driver's right/privilege to drive. Monitoring should begin with family and friends, and not wait for a run-in with the police or an accident to start the process. Handling it in stages makes the program more friendly to all parties involved. In addition, some form of arbitration and retesting should be available for a driver on the brink of losing his or her license. Obviously, self-imposed restrictions would be the best, but pride and stubbornness can often rule the day. For every person like my grandmother, who in her early seventies told the family she no longer felt comfortable driving, there are many others who continue to press their limits, steadily increasing their risk to public safety. As time passes and the percentage of older drivers grows, this could become a much more serious problem, especially when high speed gets factored in. However, many Americans are remaining active, alert and vibrant well into their late eighties and early nineties. Taking advantage of this healthy trend can also mean increased driving enjoyment—with little or no restrictions—for most older motorists.

The third group requiring improved driver education already has its program on the road. A federally sponsored mandate for the testing of tractor-trailer operators was written into law with the Commercial Motor Vehicle Safety Act of 1986. Since interstate trucking falls under federal jurisdiction, it was possible to hand down such legislation from coast-to-coast. That fact, plus the controversial enforcement procedure of depriving a state of 10 percent of its federal highway funds for failing to comply, guaranteed that this would become the first-of-its-kind national driver's licensing program. Given until October 1993 to phase in the program, each state followed uniform federal guidelines for a written and behind-the-wheel test for *all* commercial truck drivers, whether previously licensed or not. This was the first driver's test some of them had ever taken. In California, 46 percent of truckers failed the test on the first attempt. Most hadn't studied at all. With better preparation, that rate dropped to 39 percent. However fitfully, the skill level of truckers was slowly but surely being raised across the country. Once having passed the test, the truck driver's license number was entered into a national computer directory, eliminating the previous problem of bad truckers carrying several licenses from various states to hide tickets and driving records. This is a positive first step in national regulation for the safety of interstate commerce. Over the next decade, this written test and skills exam should be expanded to include mandatory training behind the wheel and a six-month, on-the-job apprenticeship to turn novice truck drivers into competent professionals. Only time will tell how such improved training and testing will relate to reduced accidents and lives saved. Three-quarters of this nation's goods are moved by truck. From the standpoint of safe, efficient commerce, it makes sense to have uniform standards that put a highly trained driver in command of a properly maintained truck moving at high speed.

Not part of driver's training as such, but vital to a driver's education, is the issue of alcohol use behind the wheel. As a nation, America has many conflicting attitudes about drinking and driving that shape our perception of the problem. Over the last fifteen years, organizations like Mothers Against Drunk Driving, the insurance industry and the various law enforcement agencies around the country have worked to pass legislation such as the national twenty-one-year-old drinking age and "zero tolerance" laws that are supposed to reduce teenage drinking and driving. Still other laws are supposed to get tough on habitual drinkers and drivers, sewing up loopholes that keep drunks off the road. Throughout America, the overriding, knee-jerk attitude is that drinking alcohol is bad. Yet, most of us do imbibe. As a result, we have created a convoluted mixture of tough talk, condemnation and either/or lawmaking that does not work.

If you look to Germany and Europe, the attitude is much different and success with lowering drunk driving much higher. There is no such thing as a twenty-one-year-old drinking law. Parents introduce their teenage children to alcohol. Wine with the family meal or a beer under the Octoberfest tent teaches responsible drinking under adult supervision. In America, alcohol is the forbidden taboo. Teenagers wind up drinking in secret anyway, with no parents to watch over them, and no one to discourage "binge" drinking. The car is still a favorite place to down some booze, in spite of "zero tolerance" laws. Europe, in general, starts enforcing drunk-driving laws at much lower blood alcohol levels than the U.S., but the fines and punishment are not that severe for the first-time offender. Yet, fatal alcohol crashes are much lower there than in America. Only 20 percent of all fatal accidents in Germany involve alcohol. In the U.S., the federal government reports that alcohol-related deaths over the last decade and a half have declined anywhere from 30 to 50 percent. Mysteriously though, the number of people killed on the road has remained virtually the same. Clearly, there is a serious discrepancy here that needs to be looked at more carefully. Now, the antidrunk-driving movement in America is pushing for a nationwide .08 blood alcohol level to determine intoxication. The insurance industry likes these ever more restrictive laws because it can raise the rates on the growing number of drivers caught in the system, whether we're having any real success with reducing alcohol-related deaths or not. MADD has grown from a small grassroots organization into a big business. Its initial pragmatism and sense of individual outrage against drunk driving have evolved into an institutional extremism that can only look at the problem in terms of absolutes, striving to reduce the threshold for drunk driving ever lower. Certain small factions within this antidrunk-driving movement view the problem of drinking as a societal evil—one that can only be solved by temperance. Prohibition was tried once in this country with disastrous results, creating a condition similar to what we had with the federal 55 mph speed limit. This puritanical streak within American culture has ultimately caused more problems than it has solved. Conversely, we don't have a good track record with drunk driving, so extreme viewpoints like this are manifested in our outrage to "get tough" with a festering problem. We will *always* be a drinking and, as time goes by, faster-driving society. Dealing more openly with both will bring more effective results than moral condemnation and fanatical devotion to repressive law and order. There is a broad range of alcohol use that finds its way on to our roads, from the occasional light social drinker to the blotto drunk—the latter requiring more serious attention. Removing the habitual drinker and driver from the road is where we have failed miserably, and where we must now place most of our effort. Individuals in the later stages of alcohol abuse have completely lost their judgment and sense of reason. They are unable to recognize the need for tough action on their problem. The majority of us as responsible social drinkers are scared to death about all the penalties we would face if caught for drunk driving. It is a very effective deterrent for us, but not for the alcoholic. Witness Clarence Busch, the man who killed the daughter of MADD founder Candy Lightner. As of the last tally, he is on his ninth DWI and still driving—even after one more serious crash with personal injuries. And so it goes . . .

A workable plan to limit "drunk" driving is not particularly difficult. It does mean adopting a new system that looks at alcohol use behind the wheel in levels of "impairment" before legal intoxication is reached. About 25 percent of all drivers in fatal crashes have some amount of alcohol in their system. The other 75 percent are stone-cold sober. For the percentage of sober drivers who currently cause these fatal crashes, all the precautionary measures discussed so far will help to reduce that number and the accidents they cause in the future. The other 25 percent can be dealt with in a progressive manner related to blood alcohol concentration (BAC). The general guidelines for this new system would take no action from .01 to .04 BAC. Between .05 and .07, the first-time offender would pay for having the vehicle towed home, but would receive no ticket or fine, just a mark on his or her driving record that would drop out of the state's computer after a year. About 2 percent of all drivers in fatal crashes are at this BAC level of "moderate" impairment. For .08 to .09, a driver pulled over for the first time would receive a ticket with points assessed, a fine and a tow charge. There would be no court case or loss of license. This infraction would stay on the driver's record for three years and be reported to his or her insurance company. Of all drivers in fatal crashes, about 1.5 percent are in this range. At .10, this would be the level of legal intoxication as it is today in most states. The gamut of court appearances, fines and loss of license would remain in place. Approximately 5 percent of all drivers in fatal crashes are between .10 and .14. For those who re-offend before the time the first citation runs out, the next higher level of punishment could be enforced. Above .14, the percentage of drivers in fatal alcohol crashes jumps up sharply to *15* percent. *Here* is where we have to start making a difference. Common sense dictates this is where we should place almost all of our effort. Striving to legislate blood alcohol concentrations for drunk driving below .10 is a waste of time. Whatever new or existing measures we use—treatment, jail or electronic monitoring—the goal should be to remove these severely impaired and habitual drunk drivers from the road.

Such individuals represent a tiny minority behind the wheel. In Minnesota, with a population of 4.5 million people, approximately 11,000 motorists have received two DWIs. This number plummets to just over 1,000 for three times and above. Though rare, some drunk drivers in my home state have been pulled over more than *twelve* times and have killed more than one person in *separate* automobile crashes! The worst time for drinking and driving is on a typical Friday or Saturday night, just around midnight. At this time, only 3 percent of all drivers on the road are over .10 BAC. Yet, they cause 46 percent of fatal crashes during that time. Only when our resources and energy are focused almost exclusively on this segment will deaths due to drunk driving fall drastically. However, this new progressive approach can also deal with those drivers who stray into the lower levels of alcohol impairment. It is vital to separate the social drinker who has had one-too-many from the habitual drunk.

The two states of "impairment" are not comparable, and the statistics back it up. Our drunk-driving laws need to reflect that fact. Otherwise, we will continue to muddle along with our efforts to reduce fatal alcohol crashes on the road.

If all of the aforementioned safety improvements are implemented on the American highway, the inevitable outcome will be more and more drivers surviving the few crashes that still happen. This opens the door for an intriguing new program of advanced driver training for crash survivors which, after fault is clearly established, retrains drivers to upgrade their level of skill and knowledge. There is no consensus to retrain every driver in America, but this is the next best thing, dealing up front with the very drivers who cause the accidents.

This program would apply only to serious, life-threatening crashes—not fender-benders—unless a driver shows a propensity to cause several minor accidents in a short period of time. Such a retraining program should not cast blame or create guilt in a driver so accused. In most cases, this person will be lucky to be alive, and this should be viewed as a golden opportunity to instill the skills and knowledge to ensure that an accident never happens again. A cost-effective system would also be a potential gold mine for insurance companies. They could offer a $1 to $2 premium increase to pay for the cost of the course if blame is affixed to one of their clients. Otherwise, the guilty party would have to pay for the course out-of-pocket. Further, these insurance companies could guarantee that premiums would not be increased after the first accident, if the driver takes the course. The benefits of this program could spread far and wide as bad drivers are given a second chance, returning to the road with improved skills, knowledge and attitude. By monitoring these same drivers over time, an entire body of statistical evidence could be collected to see if they become victims or perpetrators again. This would once and for all determine the safety and effectiveness of driver's training. We would know if the motorists who cause crashes could, in fact, be retrained. If successful, it would instill more confidence in the average motorist, helping to end the misguided belief that there are nothing but idiots out on the highway.

Several years ago, my mother accidentally let her driver's license lapse and was forced to take the basic written test—for the first time in her life, after forty years of driving. She studied hard and passed with flying colors, learning several things she never knew before. An accident retraining program would operate in much the same way, except on a much larger scale. With comprehensive classroom and behind-the-wheel instruction, drivers could learn things like skid control and accident-avoidance maneuvers. In the future, surviving an unfortunate mistake should make these retrained drivers safer, less accident-prone motorists.

The final part of this struggle to put a better driver on the road is certainly the most controversial by today's standards. We have created a curious mixture of freedom and restriction on our highways as far as the vehicles we drive. We are basically allowed to drive *anything* on the road with a minuscule amount of driver's training, but are restricted by law as to *how* we should drive. The amount of horsepower between the wheels has been a constant source of aggravation for the American highway and auto safety movement over the last forty years. Bemoaning the outrageous practice of building cars and motorcycles that can go from zero to sixty in four seconds with a top speed close to 200 mph does seem to make sense when the fastest legal limit in America is 75 mph. Given that a private pilot has to have special training to fly a plane with more than 200 horsepower, we might well question the wisdom of letting anyone with the cash climb behind the wheel of a 450 hp Dodge Viper or 513 hp Ferrari F-50. The absurdity of this contrasts well with a safety movement hell-bent on forcing Detroit to reintroduce the Model "T" Ford—with an airbag, of course.

If we are to move toward a Fast-and-Safe highway network with the possibility of a rural freeway with no speed limit at all, we must consider driver's education related to speed and horsepower. The benchmark of the whole ADEPT philosophy would be the creation of a special training program for people who want to drive high-performance vehicles with more than 250 horsepower. While an arbitrary number, this translates to an approximate value of eleven pounds of vehicle weight to every one horsepower. With these numbers as a guide, that would already put most of the motorcycles on today's market in that range and out of reach for the average rider.

The contradictions of freedom and restriction become more interesting when we switch from being allowed to drive anything at modest speed to needing special training to drive certain vehicles at any speed on the freeway. To the millions of skilled motorists who already know how to drive exceedingly well, new training requirements may seem a bit annoying and redundant. But unless your name is Petty, Unser, Andretti or Foyt, neither the other drivers on the road nor law enforcement officials know what your driving skill level is. With this advanced driver's license, there would be no guesswork.

The key to this program's success is to involve automotive enthusiasts, the racing community and law enforcement in its design and development. This coalition would put some of the best drivers in the world together to formulate an intensive course related to safety and speed under all driving conditions. With the federal government as liaison, national guidelines could be set down to make the coursework uniform and consistent from state to state. Such training must include specific programs for all types of vehicles that would be operated at high speed, from heavy trucks to exotic cars and motorcycles. Tough eligibility standards to determine who can take these advanced courses should require a minimum age of eighteen (after a provisional license is received) and a driving record clear of serious infractions such as DWI for at least three years. Health and vision restrictions should also apply, such as more extensive use of corrective lenses and screening for limitations like epilepsy. The training would then be monitored by each state police agency, so troopers patrolling our roads would feel confident about its content and implementation.

The framework of the course itself would be a minimum of twenty hours classroom and thirty hours behind-the-wheel training. Required topics would include:

- The dynamic stresses on a vehicle for normal and high-speed driving.
- Engine and chassis operation.
- Thorough understanding of the rules of the road, map reading and First Aid.
- Stress and systems management from gridlock to top speed.
- Learning to handle potential distractions on the road correctly.
- The finer points of accidents, their avoidance and reconstruction.
- A broad overview of safety and its relationship to driving and speed.

Behind-the-wheel training might not begin on public roads at all, but could take advantage of virtual reality simulation. This technology has been refined to the point where drivers can safely train within the confines of a computer-generated road course. Using a mock-up of a car with video screens in the windshield or a helmet that displays the image right before the student's eyes, a trainee can push the limits of hard cornering, fast driving or bad weather without endangering *anyone*. Already, such simulators exist for training police officers in high-speed pursuits. In the coming years, the price of such simulators will continue to drop, so even a modest advanced training program could afford it. After several hours behind a "virtual" steering wheel, a student would move on to a closed driving course. Here, he or she would work with a state trooper/instructor to learn, first-hand, an extensive array of accident-avoidance maneuvers, skid control and proper cornering techniques. Wet and dry pavement braking with and without antilock control would be a part, as well as driving on ice and snow (where climate permits). For those who imbibe, drinking and then driving on this closed course would help them better know their limits. A ride-along with an on-duty state trooper would show what law enforcement has to deal with and how that applies to the advanced driver. The course would culminate with several hours of high-speed driving on the rural interstate, taking

Lamborghini Diablo

Ferrari F355

McLaren F-1

Not just toys of the rich, exotic cars push automotive performance safety to the cutting edge, bringing technological advancement to all motor vehicles. With the proper training and safety precautions, drivers of these supercars can be allowed on our freeways at high speed, taking full advantage of their superior design— just as a 600 mph Lear jet safely shares the sky with a 100 mph Piper Cub. Photos by Guy Spangenberg cross-country trips (say 75, 100 and 200 miles in length), instilling confidence and skill while not compromising safety. The net result would be a driver's training program unrivaled anywhere in the world. Over time, as these ADEPT drivers grow in number from hundreds to thousands and, finally, millions, the comparison between these highly trained motorists and the general driving population would become a new and fascinating area of highway safety research to monitor.

Some may say that this will send an elite class of driver flying down our highways. That is exactly what is intended. If we are going to deal effectively with the prudent use of speed, a better class of driver has to be put on our roads. And this does not create a speeding, bloodthirsty monster, either. Quite the contrary. It puts an outstanding driver on our highways, one who clearly understands the conditions necessary to drive safely at very high speeds. To the average motorist worried by all this, the use of special license plates for exotic cars and motorcycles, and placards on heavy trucks would help to ease any doubts. A fast vehicle would no longer be cause for concern. Everyone on the road would know the proper precautions were being taken.

Just as fighter pilots are revered in the world of aviation, so can it be with this advanced driver program. Putting such topflight individuals on the road encourages the average motorist to "step up" to this higher level of skill. This should be exploited to its maximum potential by law enforcement and automotive enthusiasts alike, who can lead the way by taking the course themselves. Establishing ADEPT across the country will magnify the problem of nonexistent driver education in some states, and help upgrade and modify existing courses with knowledge gained in the advanced program. Over the next decade and a half, minimum standards and training would finally become a requirement for *anyone* who wants a driver's license.

Improved driver education seems like a large and almost impossible task. But it is truly a small price to pay for the privilege to drive the vast expanse of America at speeds which most of us only dream about at present, or currently experience in Europe. An agreement to bring ADEPT into being would probably be hammered out with plenty of compromise. From determining if a retest would be needed to "grandfathering in" older drivers on the maximum horsepower requirement (a similar move was made in aviation for pilots regarding the 200 hp rule in 1973), all this haggling and rulemaking would ultimately improve driving skills for all motorists. A better driver on the road will force the system to recognize the needs of the driver. Just as "Pilot in Command" caters to the aviator, "Driver in Command" would filter through our attitude about driving. Proper rules and regulations for the road would follow. The inevitable outcome of this improved training would be to give ADEPT motorists a place to go so they can *drive*—such as an interstate freeway with no speed limit.

CHAPTER 8

Interstate Patrol: A Return of Respect for the Law *and* Drivers

For all this talk of speed and safety, there is only one organization that can effectively integrate all the precautionary measures discussed and implement them on the road. It is each state's highway patrol or state police agency. No other group of men and women is more capable of showing the average motorist how to increase travel speeds on our highways while reducing the number of people who are killed and injured year after year. For the many motorists and troopers today who think of each other as the enemy, this shift in policy towards a spirit of cooperation may, at first, be a difficult transition to comprehend. Over the coming years, however, this will be the key to improving the efficiency of our road transportation system.

A special branch of the highway patrol must be developed to deal with the unique problems created by moving freeway traffic at very high speeds: The Interstate Patrol. This group would be the first to be armed with the latest high-tech systems to handle this special driving environment. As time passes, this technology would filter down to the rest of the force, but it is vital to reach the area of greatest need to get our transportation economy back on track. That means dealing with our interstate freeway system first.

As we move in this direction, it is important to recognize the detrimental effects the 55 mph National Maximum Speed Limit has had on law enforcement's attitude towards the motoring public and the public's obedience to the law itself. Over the last twenty-five years, we have become an increasingly disrespectful society in terms of the laws we set down for *all* of us to obey. This has happened on both sides of the fence, whether you are an average citizen or a law enforcement official. The "55 mph Syndrome" is largely at fault. We asked the police to enforce the unenforceable, while thinking we, as individuals, were exempt from obeying this law to the letter. Some of us began to look at the other rules and regulations on the road, feeling we were exempt from them, also. Today, running stoplights and stop signs have reached epidemic proportions in some cities of America—all part of this 55 mph Syndrome. When a majority of citizens decide that one law is not to be obeyed, over time and attrition, obedience to other laws suffer as well. It is easy to point the finger and make excuses, but sometimes it is necessary to point two at yourself to understand fully the magnitude of the problem.

I used to be one of the most flagrant violators of the 55 mph speed limit. Only on the open road, but a lawbreaker nonetheless. As I did this, I came to realize that I was as much of a problem by beating the system as those who hypocritically supported it. Such disobedience across the nation—which became so blatant that it made Prohibition look tame by comparison—has prevented us from dealing with the safety issues associated with driving. When virtually every citizen is considered a lawbreaker on the road, we become reluctant to bolster our support for law enforcement. Until we deal with the problem of reasonable and prudent speed limits for the majority, we can never make any progress with the safety considerations needed to ensure that the speed at which we voluntarily drive can also be made safer. For many Americans, moving in the direction of higher speed limits is a disturbing and dangerous step. We have allowed fear and paranoia to rule the discussion of highway safety for so long, that we perpetuate a system that punishes all of us and does not deal effectively with the problem.

To rectify this situation means pulling ourselves back a couple of notches, stepping away from the notion that "anything goes" on our roadways. We must restore our obedience for the laws of the land and return our respect to the law enforcement officials who must enforce them. In turn, those in uniform need to modify their attitude as well, treating the average motorist as an intelligent adult, not a child. Then we can begin changing how we do business on the road. First, new laws on the books must treat the majority of motorists as law-abiding drivers. Second, these laws must then be enforced by a visible, high-profile law enforcement presence. Our highways are a linear, but confined, area for law enforcement to patrol. Traveling conspicuously down the road means one trooper can cover a fair amount of ground, providing a "halo" of enforcement presence. Unfortunately, in many jurisdictions today, where speed limits are posted unrealistically low, most troopers recognize the futility of enforcing them and, therefore, do not patrol long stretches of highway. The presence of troopers now causes discomfort for most motorists, but in a higher-speed driving environment, with realistic speed limits, troopers traveling with and through the flow of traffic would elicit confidence in those using the road.

If we reinforce our commitment to law enforcement and create a special branch of interstate troopers, their needs must be taken seriously. They require the high-tech equipment to move traffic while maintaining high standards of safety. Instead of getting the automotive, computer and communications hand-me-downs, they must receive the state-of-the-art tools to get the job done properly. In recent years, the level of electronics used by troopers has been upgraded to a degree. However, we must go that extra step to provide the best technology possible. We must show ourselves and law enforcement that we are serious about dealing with the issues of safety and speed on the open road. A proper squad car, the trooper's command post, has to be the first priority for moving traffic safely at high speed.

This patrol car must take advantage of all the innovations discussed in the chapter on safe vehicle design, but with a few extras. Such a design would enhance the squad car's ability to cover vast distances quickly and safely. It would also permit the trooper behind the wheel to tap into a network of electronic information and monitoring systems to inform the driving public as to the condition and safety of the road ahead. Such a squad car would become an invaluable tool for the individual trooper, extending coverage over a longer segment of freeway.

Currently, the standard police vehicle is a slightly modified version of a full-size four-door sedan, the two most popular being the Ford Crown Victoria and Chevrolet Impala. The problem with these vehicles, and a handful of others modified for police use, is that they are adapted, in some cases, rather poorly. In the short term, they can be further refined to a degree, but a new strategy is needed to develop the patrol car of the future.

The process of building a squad car needs to be reversed. In an *Autobahn*-style driving environment, law enforcement needs a vehicle that is specifically designed for that purpose. Then, perhaps, a scaled-back version can be sold to the general public. This reversal of build technique and philosophy does not need be cost prohibitive or unprofitable, either. Basing the vehicle's design on one uniform platform—the exact same squad car sold to all police agencies—would lower build costs and maximize profit margins. With the exception of color and paint schemes for the various police forces, the patrol car would be the same, making only a small number of options available in order to keep the purchase price down. Even so, this patrol car would still cost around $35,000—almost twice what the present squad car costs when purchased through fleet sales. Even so, the extensive list of performance, passive and electronic safety innovations would make it cost-effective within the context of high-speed freeway use.

To start out with, this patrol car must achieve the best blend between all-around performance, fuel economy and pollution control, yet be dependable for mile after mile of high-speed operation. Coupling ultra-aerodynamic body work to a 3-liter V-8 engine with four valves per cylinder and electronic fuel injection fit these requirements. Developing 300 horsepower without the aid of turbochargers, this motor would be lightweight and compact, with durability and reliability to match.

Striking a balance between strength, rigidity and reduced weight, the roll-cage/frame and aerodynamics would make this squad car quick off the line and fuel efficient at high speed. Fitted with a 5-speed automatic transmission, it would sprint from zero to sixty in under six seconds with a top speed of 150 mph. More importantly, this engine, transmission and aerodynamic package would provide respectable gas mileage to keep overall operating costs and pollution low, achieving 25 mpg at a steady 80 mph—a speed that is already commonplace on today's freeway. The improved aerodynamics and more fuel-efficient engine would allow increased fuel savings for state police agencies that purchase this patrol car. Five hundred troopers driving 300 miles per day at 25 mpg instead of 20 would help to defray the increased cost of this squad car by saving the state almost $1 million per year in fuel use, not to mention the environmental savings of lower air pollution.

With four-wheel, independent suspension coupled to antilock disc brakes, this car would have speed-rated, run-flat tires, giving a firm, taut ride thanks to sway bars and antidive geometry. This suspension package could also be adapted for either a two- or four-wheel drive running gear with variable-height adjustment. Such a system would not be infinitely adjustable. Rather, it would have a more basic design that would permit a normal, low-riding position for standard freeway operation and an upper position, several inches higher, to give the extra ground clearance needed in heavy snow or pursuit driving for some off-road conditions. A small hydraulic motor could raise or lower the suspension on the fly to maximize road clearance or improve aerodynamics and handling. With all of these running-gear innovations, driving this patrol car in the speed range of 80 to 120 mph would be comfortable, effortless and extremely safe, keeping an extra reserve of speed and power for pursuit driving or emergency lifesaving missions.

The improved aerodynamics for this squad car come from the siren, megaphone and emergency running lights being flush mounted into the vehicle itself. These devices can take 5 to 15 mph off the top speed of today's patrol cars, also reducing fuel mileage. Integrating them into the bodywork means lower wind noise, better pedestrian protection and one less thing to break off or be vandalized. The dual spotlights could be retracted into the windskirts around the side-view mirrors, out of the way until needed. The flashing red lights above the windshield (blue in some states) would be the pulsing, stroboscopic variety that are more compact and maintenance free than motor-driven units. This bar of flashing lights would also be isolated from the windshield to prevent the light from "bleeding" down the edge of the glass at night, eliminating dangerous, pulsating glare. The same would be true for the rear-mounted "arrow stick" above the back window. This device can flash a brilliant yellow light several miles down the road to indicate the proper direction for traffic to move, either to the left or right, or a pulsing mode to warn that the trooper is stopped. In present-day police vehicles, these arrow sticks are quite often mounted inside, behind the rear seat, especially in unmarked cars. These bright lights can be very distracting and sometimes dangerous to the trooper. With this improved design, the standard patrol car attains the best of both worlds, being fairly recognizable as a marked squad car by day, but able to move quite covertly at night with everything tucked neatly out of the way.

As expected, performance of every part of this vehicle is tailored to maximum, safe operation whatever the conditions. From the super-bright, xenon or halogen headlight combination to a twenty-five-gallon fuel tank for a 500 mile range-of-action, every detail will have been thought out and placed into this command post for moving traffic and enforcing the law. Once seated inside, a state trooper will be in charge of an impressive array of electronic and safety innovations, some to protect the other motorists on the road, some to protect the trooper behind the wheel. The full roll-cage is made even more rigid than in the standard passenger car, thanks to the front and back seat being completely divided by a partition of steel, foam and glass. Separating the two provides a tremendous level of crash protection, especially in side impacts and rollovers. Further, it prevents suspects or prisoners from threatening or attacking the officer in the front seat. The center-divider window can be electrically raised or lowered to speak with the people in the back seat or to pass a driver's license to the front.

Since most troopers patrol alone, the driver's seat is given extra space by moving the center console slightly to the right. A passenger can still ride along in the front, but this arrangement favors the driver. Both front seats have firm side bolsters and lumbar adjustment for all-day comfort on the beat, giving outstanding lateral support in high-speed turns during pursuits. These seats are designed so accessories like gun belts and handcuffs cannot cause undue strain or fatigue and do not get hung up in the 4-point seatbelt system. The harness can also be "locked in," taking up the slack to hold the trooper in place on a hot call. No loose items, cables or cords are left dangling. Everything would have its place and be neatly stowed away until needed. Some present-day squad cars are a tangle of jury-rigged, patchwork installations to house radios, computers and ticket books, etc. Some troopers report that this is an accident waiting to happen, as anything not firmly tied down or permanently anchored goes flying in a high-speed pursuit or accident. In the patrol car of tomorrow, the center console functions as a writing table with a storage bin underneath. The radio stack, computer screen and keyboard are built right into the dashboard. The keyboard can be removed and hooked on to the steering wheel when parked to allow typing without strain, and conveniently slides into the dash, out of the way when not in use. A small video camera mounted above the rearview mirror gives an unobstructed view of the road ahead or can be turned around to view suspects in the rear seat. All critical switches (lights, sirens, ABS on/off) are right on the steering wheel to keep hands-on driving a reality while performing several duties.

One final innovation is, sadly, needed to deal with the ever increasing violent behavior of the true criminal element in our society. That is, fitting this high-tech patrol car with bullet resistant windshield and front door glass. The steel-and-foam roll-cage effectively renders the doors, firewall and partition between the front and rear seats bulletproof. Using laminated safety glass and bullet-resistant plastic, such windows would provide an extra measure of protection that law enforcement has not had. This blend of glass and plastic adds some extra weight and thickness, though recent developments have helped to reduce weight and size—at greater expense, of course. But this would be money well spent to ensure added safety under deadly circumstances. The open front door then acts as a total body shield when an officer has a gun drawn and is taking aim. Behind the wheel, an officer can pull right up to a trouble spot with enhanced safety.

This patrol car takes advantage of already existing and proven technology to increase reliability and keep build costs down. An aggressive and enterprising automobile manufacturer will, hopefully, recognize the potential boom market open to it. Such a supercar would be sought after by local as well as state police agencies. Approximately 60,000 police vehicles are purchased in the U.S. annually. That number could easily double at the outset of production of this state-of-the-art squad car. If properly designed—who knows—perhaps even the *Autobahn* police would purchase such a high-tech vehicle, along with other law enforcement agencies around the world. Extending sales to a more lucrative export market would help to reduce our trade deficit. Here, again, speed is the key. Law enforcement officers across the land have dreamed about such a squad car for years. Their continued support of underposted and restrictive speed limits is preventing them from having it. A higher-speed driving environment will help them get this technology far sooner. These facts should be an open invitation to American auto manufacturers to start building a squad car from scratch to meet the demand.

The duties of the interstate patrol to maximize safety and begin moving traffic at higher speeds would be varied and detailed. The average trooper would not be bored on the job. Moving in the direction of a no-speed-limit freeway would require troopers to drive the interstate system at extremely high speeds to uncover any dangers. Given that the original design parameters for the freeway network were between 70 to 80 mph, troopers and traffic engineers will have to locate and mark any part of the road that has line-of-sight problems or roadside obstacles that must be shielded with guardrails or energy-absorbing barriers. Poorly painted lines and confusing signs must be removed, repaired or rewritten. Speed through curves should be calculated for different types of vehicles. A small anomaly at 70 mph can become a major accident-causing flaw at 100 mph. An unrelenting, uncompromising scrutiny of the freeway by interstate patrol troopers guarantees that *they* are convinced the road is safe for high-speed travel.

This task would be accomplished by a special speed detail of the interstate patrol. Separate from, but in constant contact with, the regular troopers patrolling the road, this speed detail would operate at off-peak travel times, using aircraft and other spotters on the road to make sure no other motorists were endangered during these high speed runs down the freeway. The use of video in the patrol car can create a visual record, along with verbal input from the troopers involved. Some of the navigational and monitoring equipment from the Intelligent Transportation Systems could aid in plotting out the entire freeway network by indicating precisely where the vehicle was located at any given moment. These actions would create an extensive body of data under all driving conditions to lay out the needed improvements to the freeway in relation to speed. Over the next few years, this high-speed detail would begin to set up the support structure necessary by driving 150 mph on the freeway, so that ten years from now the general public can *move* down the interstate at 100 mph, yet maintain extremely high safety standards.

Moving in this direction of faster travel with improved safety, troopers must begin enforcing some of the new lifesaving programs on the road. Ensuring that motorists are using their seatbelts will take up a certain amount of time. However, a majority of motorists already buckle up and compliance should jump noticeably after higher speeds and, hopefully, tougher seatbelt laws go into effect, so this enforcement should not be all consuming. The implementation of the Drive Right Except to Pass program will be a more involved task. This breaks down into three areas of driving technique, education and enforcement. The Drive Right Sweep, as previously discussed, helps to align traffic in relationship to lane and speed. Using the left turn signal and flashing headlights to indicate that they would like to pass, troopers driving up the left lane above the flow speed of traffic would move slowpokes over to the right lane. For Left-Lane Bandits and fast drivers trying to pass on the right, being pulled over by a trooper becomes an "Educational Stop," offering information on the safety and effectiveness of the Drive Right law. It is important, whenever possible, that troopers themselves move over into the right lane, indicating that Drive Right is for everyone.

Along these same lines, having troopers drive in the right lane *under* the flow speed of traffic will get motorists used to the idea of passing the police. For many Americans, this can be disconcerting. The first few times I passed the *Autobahn* police were a bit unnerving. Such a Drive Right Pass on the American interstate would help to set drivers more at ease, until overtaking a squad car becomes no more unusual than a trooper passing them.

As expected, a faster highway would mean a more rigorous crackdown on impaired driving and, especially, habitual drunk driving. With traffic speeds increasing, any person under the influence becomes a greater accident risk. Given higher speed limits and, finally, no speed limit at all, the average driver will no longer be hunted down by the patrol. Enforcement can then shift to the more serious problem of drunk driving with the average motorist aiding in that goal. With all of the cell phones in cars, the previously discussed 911-like number direct to the state patrol would put drivers and troopers in contact with each other to reduce this problem. This number could also be used to report accidents, breakdowns or criminal behavior other than drunk driving.

Tomorrow's high-profile trooper must base all duties on the premise of maintaining the highest level of safety within the fastest flow of traffic under varying degrees of weather and road conditions. This means the speed/safety relationship could undergo a great deal of change during one eight-hour shift. Road construction, rush-hour traffic, a change in weather and the aftermath of a minor or serious accident could create many variables the interstate patrol would have to deal with. The implementation of the Advanced Traveler Information System is vital to the improvement of freeway safety in the coming decade. A trooper hiding in the bushes or on top of a highway overpass just waiting for trouble cannot effectively maintain a safe flow of traffic. Moving down the road keeps the trooper alert to problems that must be relayed to all other drivers via the ATIS network. At high speed (near the 100 mph mark), it is possible to cover tremendous distance in a short period of time. Knowing "what's up ahead" becomes important for everyone behind the wheel. ATIS also indicates to motorists that the patrol is out there on the road, overseeing the safe, efficient movement of traffic. Law enforcement's presence is "felt" every time new information comes over the system.

As such innovations come into widespread use, their control, manipulation and proper function must be watched over by the interstate patrol. Devices like line-of-sight warning indicators and the Variable Speed Limit System—whether in rural or urban settings—are under normal circumstances computer-controlled and operate automatically. Under special conditions such as an accident or stranded vehicle, the trooper must be able to take manual control of the system by tapping into it from the computer console in the patrol car. Keeping the automated part of the freeway under hands-on human control will maximize safety and efficiency. However, it is also vital to have checks and balances to make sure the human element does not forget to return the system back to automatic control after a problem is corrected.

With this increased technology and roadside safety improvements, troopers should work in tandem with one another on both sides of the interstate. The coming use of guardrails and antiglare fences to protect the center median strip effectively prevents anyone from seeing over to the other side. One trooper patrolling each direction of travel while in contact with the other can better monitor a smooth flow of traffic. Under serious or life-threatening conditions, backup is only moments away. Troopers can no longer jump the center median, but with the superior design of the squad car and the high-speed nature of the road, they can make the next on/off ramp in record time to aid a partner in trouble.

Such backup for troopers should be further expanded into a much broader electronic support structure across each state and the entire country. State and national databases on criminal activity would be only moments away on the Internet. Transmitting audio and video from car-to-car and station-to-car would aid in the apprehension and prosecution of wrongdoers. Starting in our larger cities, transponder-equipped patrol cars could be monitored as to their location and present duty. This system could become a statewide grid over time. Metropolitan areas with large volumes of freeway traffic could then more effectively allocate troopers where they are needed most. Once such an electronic network were extended over a state's highways, especially in the larger western states, even troopers isolated by distance would have sufficient electronic support.

Above and beyond these ongoing duties of safety, traffic control and law enforcement, the state patrol along with traffic engineers will have to monitor the trend of increasing speeds on our rural highways and freeways. On the open road, speeds have continued to creep up as time passes—about a 1/2 mph per year. Recording these speed trends and adhering to the "5/95 rule," will be a new priority for the state patrol. Traffic engineers would then base highway speed limits on a minimum set at the 15th percentile speed and a maximum at the 85th. Enforcement would begin at the 5th and 95th percentile marks respectively.

These criteria developed out of studies undertaken in the 1960s and early 70s by David Solomon and Julie Cirillo, and out of traffic-engineering research by Leonard West and J.W. Dunn. Unfortunately, our response to the Arab oil embargo, the 55 mph National Maximum Speed Limit, and the subsequent Slow-is-Always-Safer movement, hindered implementation of the 5/95 rule throughout America. At first glance, speed enforcement at these percentile levels sounds extreme, but examining the speed data for several segments of rural interstate gives a different impression. In general, and taking into account a few miles-per-hour variation for the region of the country and the free flow of traffic, the 85th percentile speed on the rural freeway is approaching 78 mph, the 15th percentile speed around 60 mph. Moving up and down the scale, the 5th percentile speed is about 55 mph and the 95th percentile speed is around 85 mph.

Using this data as a guideline, the maximum speed limit for today's rural interstate should be 80 mph with a 60 mph minimum. Law enforcement would begin to target exceptionally slow drivers around 50 mph, fast drivers about 90 mph. For trucks and buses unable to maintain the minimum speed because of steep grades or heavy loads, a requirement to remain in the right lane and use their emergency flashers would indicate to faster motorists that a slow-moving vehicle is ahead. This is already law in Pennsylvania and other mountainous states and should be extended to the entire interstate network as well. Such a mandate could also be required for passenger cars, but such slow speeds often indicate timidity and lack of skill on the part of the driver. The previously mentioned traffic safety studies indicate that these slow vehicles represent a greater accident risk, more so than extremely fast ones. Here is where the slow-driving equals safe-driving myth is exposed. The interstate was built for speed, and this tiny minority of motorists should be required to leave the freeway and use slower secondary roads.

Tying speed enforcement to the 5/95 rule will keep the safety/speed relationship out in the open. When flow speeds of traffic begin to extend beyond the posted limit into the grace given by law enforcement, the limit should be raised another 5 mph to accommodate this natural trend among the majority of drivers using the road. It remains to be seen if this gradual increase in speed will plateau at some voluntary maximum due to law enforcement presence, driver behavior, and/or road and vehicle design. While we will never have enough police to enforce speed limits strictly on all our roads, general societal consensus in concert with a high-profile patrol might bring about a voluntary maximum for our rural highways. These speed limits would be quite high by today's standards. But if all the other precautions of driver, vehicle and road design are phased in, they can be handled safely.

By the end of the next decade, exceptions might be sought for the rural interstate system in those states that want to move beyond the traditional notion of a speed limit, pursuing an *Autobahn*-style freeway with no speed limit at all. The conversion should occur when voluntary flow speeds for passenger cars and motorcycles reach 80 to 90 mph. Most states and, in fact, most countries except Germany do not have the vision to deal with the safety issues of speeds this fast. Within this 80 to 90 mph range, a state must make the commitment to an unrestricted-speed freeway to maximize its potential. At these speeds, a support structure based on the concept of fast driving with very high safety standards must take shape. The preparation of the road, vehicle and driver along with law enforcement presence come together at this point. The benchmark for the freeway then becomes: Drive at a speed at which you feel comfortable, drive with respect to the prevailing road and traffic conditions. Continuing to monitor rural freeway speeds after achieving an unrestricted-speed designation will guarantee that further precautionary measures will find their way to the interstate in a timely fashion.

Despite all of these improvements in safety, law enforcement will have to deal with the fact that some accidents are still going to happen. On higher-speed roads, Drive Right, the Variable Speed Limit System and removal of drunk drivers will help to reduce the frequency of accidents. When a collision does occur, improved vehicle design, a high percentage of seatbelt use and upgrading roadside safety will lessen crash severity. However, a faster flow of traffic will create its own unique accident problems whether it be single-car crashes or multivehicle pile-ups. When law enforcement is prepared for every possible contingency, the effects of speed—and its by-product, the sudden deceleration in crashes—are minimized. The continued use and expansion of air ambulances to quickly remove injured people to the hospital will help lower the death toll. Clearing accidents as soon as possible to prevent second and third "gawker" crashes will be more urgent as speed increases. Here, ATIS and electronic road signs can quickly provide alternative route information to shift traffic away from the problem, or prepare those in the immediate area to slow down. Motorist's education by the patrol becomes more critical as the potential of people milling about a smashup or breakdown will have more serious implications with traffic whizzing by at high speed. If people know it is safer to stay inside the vehicle or move as far off the shoulder as possible, they will refrain from standing on or walking down the freeway. Something as simple as having troopers stay with a stranded motorist or at a crash scene until it is completely cleared could spell the difference between a minor mishap and fatalities. Basing law enforcement procedures on "if it can happen, it probably will" means being prepared for anything.

Other duties will expand the role of the interstate patrol and, more broadly, the entire state police agency. The improvement or creation of vehicle inspection programs and the patrol's participation in the development of the various ADEPT driving courses will work to ensure that what is happening off the road brings the greatest safety to the road. Such "hands-on" monitoring and education will create a better working rapport between law enforcement and citizens behind the wheel. Greater confidence will develop in both as precautionary measures are phased in. A sense of common purpose will begin to filter through our attitude about driving and safety. However, there will always remain a certain barrier between those working in law enforcement and the public at large. The average citizen will probably never think of the average cop as a buddy or friend, but striving for similar goals on the road should achieve more respect and obedience for the law.

One final job the state patrol manages has not been exploited to its full advantage. That is the investigation and reconstruction of accidents to save lives. This science has been refined to quite a high degree, in courses offered to law enforcement and traffic engineering students at several universities across the country. Already, reconstruction of accidents is used extensively in court cases to determine who is at fault. In the areas of improved vehicle design, roadside safety and accident reduction, the use of reconstruction can be expanded. *Every* trooper on the road must be properly trained to record accidents. The first officer on the scene would then be instantly ready to measure, videotape and photograph the crash site without missing valuable evidence or holding up traffic that can quickly back up for miles. "Black boxes" to record speed, direction and force of impact are already on the road in a growing number of vehicles. These devices might also speed up and enhance the recontruction process. Once collected, the data should then be placed into a national computer network for other police and safety agencies to draw from and compare with similar accidents. In a Fast-and-Safe driving environment, it is necessary to assess this data with regard to minimizing the risks of speed. The same approach is used in aviation crashes to supply meaningful change to lessen the dangers of flying. Speed is a given, just as it should be for the interstate system. The information gleaned from this process can then be used to bring the best educational and technological advancement to the road. Improvements in accident investigation could extend beyond the physical site of the crash. By interviewing the survivors and their families, a growing body of psychological evidence could profile those involved in serious and fatal crashes. We already know a great deal about accident risk related to the age and sex of drivers, but almost nothing about how risk correlates with background and mental state. Hopefully, a certain amount of knowledge will come from such profiles and improve safety by informing the driving public about those at risk.

A high-tech state patrol overseeing a high-speed interstate offers many positive benefits to both law enforcement and the driving public. The patrol gets the technology to deal with the special considerations of faster driving and saving lives. Motorists see a movement away from writing tickets that specifically targets them, toward traffic management designed to help them safely reach their destinations. The patrol can tap into a whole new driving environment where obedience to the law is the rule, not the exception. They would no longer be tied to a quota system of speeding tickets. The three "Cs" of enforcement—Courtesy, Courtesy and Courtesy—will return again to the American road. The highway patrol can go back to its roots, once again becoming a service-oriented organization that also enforces the law, not the other way around. As a result, respect for the law and those enforcing it will become more prevalent. Driving faster will cause most motorists to pay more attention to how they drive. These same motorists will also be more sympathetic to the financial considerations needed for law enforcement to monitor traffic safety at high speed. Any state interested in raising its speed limits further should ante up to pay for its patrol to participate in the Fast-and-Safe agenda. Just one cent per gallon on the gas tax would cover the cost for most states, providing more than enough money for improved technology and increased personnel. However, it is also important to remember that you don't need a trooper at every highway overpass when the vast majority of drivers are voluntarily obeying the rules of the road.

As we begin to post higher speed limits and travel faster on the open road, there will be a whole new body of data to monitor and analyze. We can begin to show ourselves, once and for all, the safety implications of treating ourselves as law-abiders, not lawbreakers.

However, any endeavor we undertake has both expected and unexpected ramifications. In many respects, driving faster becomes a national experiment on ourselves. To guarantee success, we must involve not only the average motorist, but state patrol and state police agencies across the land.

CHAPTER 9

A Drive on Tomorrow's Freeway

Here is the payoff for a job well done, the reward for making the assumption that high speed will be a part of our future highway safety network. From the layout of the road to the rules of the road, from vehicle design and inspection to driver's training and attitude, nothing will be left to chance with all of these precautions overseen by an interstate patrol monitoring tomorrow's Fast-and-Safe freeway.

It's time to take a drive and see for yourself what all our hard work has accomplished.

The first thing you notice is how *red* it is. With the targa top open, it looks low and squat, hugging the ground with rounded curves and flowing lines that speak an unmistakable tradition: Corvette. This one is the fiftieth anniversary edition complete with speed-sensitive suspension and the classic 350 V-8 with 400 horsepower. In the past, there have been Vettes with bigger engines and more power, but none as lightweight, aerodynamic and fuel efficient. Nothing has been overlooked. From the factory sponsored driving course to the roll-cage with certified crash rating, this is the ultimate high-speed machine for America's interstate.

As we walk up to the car and give the tires a quick once-over, you wonder how you will manage to scrunch down and get into something so low. When you open the door, however, the seat is facing you and it is a simple matter to sit down and twist your body and the seat into position. The doors close with a reassuring "thud," locking themselves firmly into place. It's hard to believe that doors so strong and thick could be so lightweight and easy to close. With almost four inches of padding on the inside, and a web of fiberglass, steel and foam to protect you on the outside, the entire door becomes a part of the car. This is not just to save you from someone running a red light, but to give the car the strongest possible frame for hard cornering and handling at high speed.

(I know, you're itchy to get going, but there are a few other things to tend to first.)

The whir of a motor deep in the seat brings the three-inch-wide, 4-point seatbelt over your shoulders. Slipping your arms underneath, you lock the harness into place. A slight tug on each shoulder strap takes the slack out of the belts and now you and the car are one. A small lever on the side of the seat ever so gently puts a light tension on the seatbelt so you are snugly held in place. Remember, this *is* going to be a fast drive. If you need to, a simple flip of the lever keeps you strapped in but free to move around. Otherwise, staying "locked in" gives that extra edge if we should get into trouble.

Everything feels right and within easy reach. From the analog gauge package behind the steering wheel to the 6-speed shift lever on the center console, the layout is simple and complete. The radio/CD player is something different, though. It's really two radios in one. The normal AM/FM and a second FM receiver for the Advanced Traveler Information Service, giving the latest reports on interstate road and weather conditions. A flip of the switch to ATIS along with the correct frequency for our stretch of freeway—and . . . We have to wait a few seconds for the reports of other local sections to finish. Today, we could be driving from Minneapolis to Chicago, San Francisco to L.A. or Louisville to Atlanta. Finally, the report for our segment of freeway comes up:

> Interstate patrol weather and freeway conditions with information *A*lpha: Clear and sunny throughout route length. Winds from the west-northwest at 3 to 5 mph. Temperature 63°. Freeway road conditions indicated as excellent, except for one five-mile section southbound down to one lane—shoulder and right lane under construction—starting at mile marker 115: speed limit 50 mph. Caravan of thirty antique automobiles entering freeway system near mile marker 175, southbound in right lane from 10 a.m. to exit near mile marker 225. Estimated to leave freeway system at 10:50: tour speed 60 mph. Light to moderate truck traffic along entire route, speeds of 70 to 80 mph reported. Watch warning signs for trucks passing on hills and curves.

That's it. Time to get moving. Everything's checked and double-checked: seatbelts, mirrors, controls. A twist of the key and the engine comes to life. A spirited, throaty sound of precision machinery you can feel as you press down on the accelerator and pull away from the curb. The Vette surges out onto the street. It's a beautiful May morning and the kids on the sidewalk are just heading off to school—eyes glued on the red Corvette as we putter by. This is the place to hold it down and take your time. We're still a couple miles to the entrance of the freeway so we leave the ATIS rambling on to listen another time for our stretch of interstate. We'll have no need to constantly monitor the radio as we drive down the freeway. The letter "A" is prominently displayed next to the clock. And in tune with the Corvette's state-of-the-art nature, our radio has ATIS plus—automatically turning on when new or emergency information comes in from the interstate patrol.

There it is, just up ahead. The familiar red, white and blue crest of the interstate system, an arrow pointing to the right towards the entrance ramp. It's all business now. The crisp morning air in our faces gives way to the electric whir of the targa top and windows closing. Instant silence except for the clicking of the turn signal and muffled hum of the motor. Now, the engine comes to life as we briskly move down the acceleration lane and onto the freeway, giving us a push into our seats as we quickly move along with the flow of traffic.

The speed of the traffic is what you notice first. You're a bit nonplused that so many vehicles can be moving so fast on three lanes of freeway. The second thing that stands out is how everything seems to be in such sharp focus. The markings on the road, the paint, signs and guardrails are perfect, clean, easy to read and straightforward. The first variable speed-limit sign on the overpass above indicates a 60 mph minimum and an 80 mph maximum. Slower vehicles to the right, faster to the left. When a break between the cars comes, we ease over two lanes. After all, you *are* a fast driver.

Traffic is fairly heavy this Tuesday morning, moving just under the posted limits, but it is a marvel to slip past all these gleaming, low-slung cars—a fast-paced flow of bright colors with a hint of chrome. The ultra-aerodynamic 18-wheelers move smartly down the right and center lanes, their windskirts covering all but the front wheels, their huge federal inspection stickers with driver's license number prominently displayed on the back end. Doesn't anyone drive a dirty car these days?

As the traffic begins to thin out and the speed picks up slightly, you notice a couple of other things. Everyone, and I mean *everyone*, is wearing a seatbelt. Whether it's out of fear, perceived danger, common sense or the law, every driver and passenger is strapped in. The other thing is the interstate patrol. You "feel" its presence. It's not that there are squad cars around every bend in the road, but you see them regularly moving with the flow of traffic. In fact, there's one just up ahead. We're steadily gaining on an interstate trooper as he flows along with the center lane of traffic at a sedate 70 mph. No longer a big, boxy patrol car, this Chrysler police pursuit cruiser is so low and sleek—all decked out in state colors and crest—that you almost have to look down at him as we move on by. We hesitate for a moment in passing him (old habits die hard). No matter. He has no more concern on his face than if he had passed us.

Now the fun begins. We've quickly moved out of the main stream of city traffic and are approaching the open road. The last few variable speed-limit signs have indicated the same maximum and soon—look—the sign all fast drivers long to see: the gray circle with diagonal lines through it—the end of speed restrictions.

Passing underneath the last variable speed-limit sign and checking to see that the ATIS is still on information Alpha, it's time to open up the Corvette. 90 . . . 95 . . . 100. It's amazing how smooth and effortless it is to drive this car at these speeds. No wind noise, the suspension absorbing any irregularities in the road. The car surges forward as the long, flowing curves in the freeway seem to be eaten up by the front end, inhaling the pavement. A pleasing rhythm develops between the car, the road and you. Driving Right, Except to Pass, you flow by the other traffic with a fluid, triple glance in the left-side mirror. This allows you to gauge the speed of vehicles behind you to see if they are closing in quicker on you than on the car you wish to overtake. Today, we're moving about as fast as anyone else on the road, so no one, as yet, has rocketed up from behind to pass.

There's good spacing between the cars and trucks moving along in the right lane between 80 and 90 mph. Overtaking them at 15 to 20 mph faster poses no problem. The occasional car that does pull out to pass gives us plenty of room to ease off the gas, allowing them to safely overtake a car, truck or motorcycle. Today is one of the better days for this kind of fast driving. Later in the summer, the vacation traffic builds to a more hectic pace, making these kinds of speeds impossible. But at the moment, the stream of traffic is stretched out uniformly, not like the old days when blanket speed limits bunched up cars and trucks into caravans. Gone, too, are the Left-Lane Bandits, blocking traffic by loafing in the passing lane. This also helps to spread out traffic, making higher speeds possible. There *is* the occasional slow driver, puttering over in the right lane at a reserved 60 mph . . .

A little less than an hour has gone by, and the mile markers are quickly approaching 115. Just like clockwork, the interstate patrol appears, the cruiser's rear-mounted arrow stick pulsing its bright yellow lights toward the left lane. Speeds drop quickly and the cars and trucks bunch up bumper-to-bumper as the traffic merges into the left lane for the coming road construction. The number of large, portable flashing arrows and signs makes the freeway look like the entrance to a casino in Las Vegas—no sense taking any chances at these speeds.

The speedometer needle flutters either side of 50 mph, and after an hour of driving twice that, we seem to be crawling along at a snail's pace. We dutifully chug our way down the one-lane freeway, glancing over to the other side of a temporary cement guardrail to see the road crew resurfacing the right lane and shoulder of the interstate. A series of lights and reflectors on either side of us clearly outlines the road—especially at night.

This is as good a time as any to catch up on the latest news on the radio. Another banner year for the U.S. auto industry, fifth straight year of record sales and growth. Boom times in the Motor City. Last year's highway death statistics are also out on the heels of all this economic prosperity. Twenty-five thousand people lost their lives last year, just under 3,000 on the interstate system. Still too many, but this is the tenth year in a row of decline. There are now too many zeros after the decimal point to talk about the death rate anymore. The percentage of decline in the number of deaths from year to year is given. Last year: down 12 percent.

It's only been a few minutes delay, but this slowdown for road construction seems to be taking forever. Finally, the speed picks up slightly and the traffic moves apart a bit. Suddenly, we're off and running again. Like water seeking its own level, the traffic quickly thins out and finds its own speed. The big trucks promptly move over to the right lane as the cars leapfrog one another left-right, left-right, spreading out down the road. In less than a minute, everything is back to normal and the Corvette is cruising once again at just over the 100 mph mark. The exhilarating rhythm of the road returns. The minutes and the miles click by in double time.

Just a few minutes late, but more or less on schedule, is the caravan of old cars entering the freeway. Parked near the entrance ramp is an interstate patrol car with yellow lights flashing to make sure the group gets safely on the road. The faster cars and trucks flow over from the right lane to the left as these former Kings of the Road merge onto the interstate. Briskly moving past Hudsons, Studebakers, old Chryslers and Fords, we're at the head of the pack and don't get bogged down in this temporary traffic jam. This leaves us free to take advantage of the empty road ahead.

Putting the pedal down to see what the Corvette can do, we start up a steady rise, but flashing yellow lights are, again, pulsing in front of us. We've just come across one of the few line-of-sight warning indicators on the freeway. Backing off the gas and touching the brake—ready to step on it if need be—we crest the hill in plenty of time to see one semi overtaking another at about 50 mph. Had the warning device not been there, we still would have had enough time to stop with the Vette's antilock brakes, but it would have been close. These warning systems that sense speed on both sides of a hill or curve have been lifesavers on the interstate, preventing scores of rear-end accidents at high speed, especially at night and in bad weather.

The one tractor-trailer slowly moves by the other and eases over into the right lane, revealing a wide-open, empty stretch of freeway. It is a timeless moment. One that calls for the best that Beethoven can provide, perhaps the Fourth, Seventh or Ninth. A dropped gear and the accelerator pedal to the floor, we surge past the trucks and down the open road. The CD player rings out as the Corvette responds like a thoroughbred bolting from the starting gate. The speed builds quickly, 110 . . . 120 . . . 130. A slight falling off of acceleration as the center guardrail and antiglare fence stream by in a solid flow of green and silver. The road ahead is crystal clear as the shoulders blur by in a slowly undulating wave from the subtle tucks and rolls in the road. It feels fast, *very* fast. The speedometer needle drifts ever so slightly past 140. The road ahead is still clear. Thirty seconds become a minute. We could go faster. The road slopes downward, allowing us to see several miles ahead. We're traveling the length of a football field every 1.5 seconds, 2.3 miles every minute.

I'll bet you nobody's going faster than we are . . .

Headlights in the rearview mirror. They're well in back of us but closing fast. Whoever it is has the left turn signal on and wants to pass. We're already in the right lane and taking no chances. The car draws even for a split second then bursts on down the road. A woman about forty years old driving a steel-blue Lincoln Mark X, the new one with the high-output, twin-turbo charged engine. Gets 40 mpg at 70 mph. Today, she's well over twice that speed—a lot of high mileage and horsepower for a $60,000 supercar. This isn't a race. Even though the Corvette is slightly faster than the Lincoln, 140 seems plenty fast enough for today. She must be late for a business meeting . . .

Seconds pass, the Lincoln is now just a pinprick of black as sporadic traffic appears on the horizon. We ease back near the more comfortable 100 mph mark and slip into the flow of traffic. The music hits a forté then vanishes into an electric silence that is replaced by the familiar high-pitched beep-beep-beep of the ATIS alert: a stalled car just up ahead. We quickly signal and move into the right lane as an interstate trooper zooms up from behind us in the left lane. The cruiser's headlights pulse high-low, high-low. Its flashing red lights shift to brilliant yellow as this trooper streaks past us with her low-riding squad car. She synchronizes up with the traffic, crosses over in front of us and rides the shoulder up to the stranded vehicle. Coming to a stop behind it, the patrol car raises up several inches on its suspension (the off-road mode), giving the cruiser the highest possible profile for its flashing yellow lights. The four people in a stalled Toyota dutifully remain strapped into their seats as the trooper gets out of her car to help them.

I hit the ATIS reset button and Beethoven returns in all his glory as our speed picks up again to triple digits. The radio flashes a new "B" for information ***B***ravo. Out of the corner of your eye, you catch the flashing lights of a tow truck through the green antiglare fence as it comes down the other side of the freeway to pick up the stranded car. Our road rhythm returns and we press on with a muted sense of exhilaration. We're not tired, certainly not bored. Just keyed up with the high tension of driving this long, this fast. It helps you to focus your concentration on a demanding, but enjoyable task.

The last few miles, the traffic has been slowly building, the distance between the cars and trucks drawing closer. Our speed drops below 100 mph. This transition area between low and high density traffic is where most of today's freeway accidents have been happening. That is why in many larger metropolitan areas, the Variable Speed Limit System has been extended out into short stretches of rural interstate. The first 80 mph variable sign is just up ahead. It seems slow but not so slow as to be unreasonable.

Traffic is building quickly and we should get over to make our exit. Off in the distance, the outline of the city is highlighted in blue sky and puffy, white clouds. A high-speed train flies by to the right, carrying noon-day commuters downtown. Our speed drops below 40 mph as we edge over into the right lane and prepare to exit the now swollen freeway. Up the ramp and to the right, we'll take a moment and pull into the nearest gas station to fill up.

There's an unfamiliar sign below the station marquee, a bright yellow sun with an electric plug coming out of it: a battery exchange and recharge station for electric vehicles. Once you know what to look for, you see a few diesel-electric hybrid vehicles and even the occasional electric commuter car out on the street. As we pull up to the premium-unleaded pump, your eyes bug out at the $2.75 price for a gallon of gas. You get to fill up. I'll pay for it. The digital numbers pulse by but the Corvette takes only twenty-two gallons. Not bad when you think we just covered 400 miles in under four hours. Just a short ten years ago, it would have taken more than six hours for the same trip. Now, after lunch *I* get to drive back and maybe this time we'll get to find out how fast that Corvette really goes . . .

An impossible dream? Utter madness?

The fact of the matter is, this is happening right now. Not from Minneapolis to Chicago, San Francisco to L.A. or Louisville to Atlanta; but Munich to Stuttgart, Frankfurt to Hamburg, Hannover to Passau—without as many safety precautions *and* at even higher speeds than we just drove on tomorrow's interstate.

One thing is clear. Fast and Safe can coexist on our freeways. The only thing lacking is the drive to make it happen here in America.

CHAPTER 10

Come American Autobahn?

Will there ever be a rural American interstate with no speed limit?

The answer is absolutely *yes*. Or it could be absolutely no, depending on what direction we choose to take as a nation from this point forward. When it comes down to the options before us, only three possibilities remain: The first is to do nothing, maintaining a status quo which all but guarantees that 40,000 people will die this year and every year afterwards. The second is to make a true commitment to the Slow-is-Always-Safer credo by allowing safety experts and Washington bureaucrats to determine what constitutes a slow enough, "safe" speed, and then enforcing this limit by retrofitting *every* vehicle on the road with mandatory speed governors—highly unlikely after the repeal of the 55 mph speed limit. The third, and final, option is to work to minimize the accident risk of higher-speed driving and to integrate this Fast-and-Safe philosophy into our highway safety network.

The first option of doing nothing is not acceptable. To throw up our hands and say that this is the best that we can do, that 40,000 people dying annually is tolerable, cuts against the grain of what this nation is supposed to stand for. Likewise, an America that measures distances in thousands of miles and was founded on the principles of freedom and the pursuit of happiness will continue to resist being forced by its government to slow down—the twenty-one years of federal 55 mph speed control have shown that to be true. But with freedom comes responsibility *and* self-sacrifice. Working to achieve a faster, safer American highway is the "Good Fight," but it also means taking the more difficult road to improve a challenging and complex problem which has plagued this nation for more than 100 years. Such a call to automotive arms is nothing new. Building the consensus necessary to implement successfully the Fast-and-Safe agenda on the road will be a challenge as difficult as any this nation has ever faced. Our ability to unify against the common enemy of Death on our Roadways is as great a test of our national character and strength of will as is the inevitable temptation to declare war on ourselves in the battle to resolve this problem. To paraphrase Lincoln: A nation against itself cannot succeed. When the government works against its own citizens and traps law enforcement in between, we can never hope to make any real, long-term progress with saving lives on the road. When the American automotive manufacturers are unwilling to make the best cars in the world, and we as a nation abdicate our responsibility to guide them in the right direction, we lose on two fronts: The bitter reality of jobs and prosperity lost to foreign competition, and the reduced transportation efficiency that saps our economic vitality through slow speeds and thousands of annual traffic deaths. This divided-house mentality was tolerated with a certain sense of detached bemusement when the threat of foreign competition and the magnitude of death and suffering on the road were not clearly understood. No one can remain so cavalier today. As the search for common ground grows ever more frantic, Americans must unify to solve our problems. No other way is possible if we are to evolve into a better and greater nation.

The leaders of this movement to guide the government, the automobile manufacturers and the driving public in the right direction must be those of us who not only like cars, but who *love* them. The burden of proof to show the nation that Fast and Safe can work rests with those of us who wish to drive faster. The millions of automotive enthusiasts across this country are responsible for the future development of performance and passive safety in the American automobile and on the American road. No other group of men and women is more qualified for the task of influencing our national agenda on these matters. The time has come for those of us who love to drive to have a say in *how* and *what* we drive.

The first step of this challenge will be to move the nation in a more progressive direction toward safety and speed, the top priority being a set of precautionary measures to ensure a reduction in deaths and injuries on the road. States interested in pursuing the Fast-and-Safe agenda should adopt the following guidelines:

1. Mandatory seatbelt-use law with primary enforcement, a $50 fine and points assessed on a driver's record for failing to comply.
2. Progressive system of blood alcohol concentration for "impaired" driving, starting at .05. The development and implementation of a comprehensive plan with fines, treatment and punishment to remove habitual drunk drivers above .10 blood alcohol level from the road.
3. A Drive Right, Except to Pass law for applicable stretches of interstate/limited-access freeway.
4. Five-cent per gallon gasoline tax to fund roadside improvements for safer, higher-speed driving. Two cents for freeways, two cents for two- and four-lane highways and one cent for law enforcement agencies.
5. A law requiring speed limits to be set at the 85th percentile and a more extensive use of 15th percentile minimum limits. Enforcement of the "5/95" rule for these new speed limits.

This five-point guideline program gets us off to a positive start. But when setting out to accomplish any difficult task, it is important to have a national goal to strive for. The battle cry for the American highway safety movement should become 20 by 10, working to reduce the overall number of fatalities on the road from 40,000 to 20,000 by the year 2010. Since speed limits on our rural highways and freeways could be raised during that time from 65-75 mph to 80 mph and beyond, this presents a complex challenge which will require innovative action at many different levels. In the short term, it requires the rules of the road to reflect the needs of drivers who use the system the most, making the occasional user step up to this higher level of skill. We can no longer cater to the lowest common denominator by tolerating all kinds of nonsense. Making modest improvements in proper lane obedience, realistic minimum and maximum speed limits and the return of simple common courtesy is the only way to improve our collective driving ability. Here again, car lovers leading the way to a shift in law enforcement tactics, use of signs on the road and positive educational information in the media can create an environment conducive to better driving habits. Longer term, the driver's training agenda must set down national guidelines for content and a minimum amount of classroom and behind-the-wheel training for *anyone* who wants a driver's license. This program must also work to establish a minimum driving age of eighteen. Taking into account the diversity of human nature, this long-term improvement in driver's education still has the potential to put a better driver on the road who has the skills and attitude for a lifetime of safe driving. This does not mean that everyone on the road has to pass an Indy 500 driver's test. What it does mean is an incremental improvement in skill, a high degree of voluntary compliance with a handful of crystal-clear, rigidly obeyed laws *and* the abandonment of the false and divisive belief that everyone on the road is an idiot. (Remember, they're thinking the same thing about *you* and that guarantees nothing will ever change.)

This shift in policy to bring about better driving ability, road conditions and law enforcement practices must then involve the automobile manufacturers. Rewriting the minimum crash protection standards for front-offset, side-, rear-end and rollover crashes would bring superior roll-cage design and improved seatbelts to the marketplace in a more timely fashion. It should also leave the car companies free to design in the most cost-effective modifications without mettlesome micromanaging from Washington bureaucrats. Less expensive, all-around safer vehicles would follow. *Performance* safety improvements should include industrywide use of speed-rated tires, an ABS on/off switch and more extensive advertising to educate the motoring public on the proper use of antilock brakes. Such changes over the next few years would put us well on our way to meeting the 20 by 10 goal.

As the symbiotic relationship of faster driving develops a need for more performance and passive safety, the opportunity exists to create an entire revolution in automotive safety design. If launched by the American automobile manufacturers, this trend could propel consumer demand for these innovations into a period of increased sales and economic prosperity. This revolution could be as exciting and prosperous as when car styling of the 1950s brought boom times to Detroit and the rest of America. It requires the U.S. auto industry to act *now*. Time is a precious commodity and the threat of foreign competition paying more attention to this trend than does Detroit could set off another period of American economic decline. We can remain cautiously optimistic, however. It appears that the Big Three desire to remain so, opening the door to a host of improvements, ranging from antispool seatbelts to rupture-resistant fuel tanks. Working from the premise that the comparable American car still costs less than the foreign competition, domestic auto manufacturers can build in more performance and passive safety, yet price their cars lower, thereby making up the per-unit profit difference through increased sales. It's a tricky game to play, but a challenge worthy of Detroit.

If this trend in vehicle design and all other highway safety improvements is established, the number of people killed on America's roads could be lowered to under 10,000 by the year 2020 even though rural freeway speeds of 100 mph, or more, would be commonplace. As always in this endeavor, no other area of safety development is as important as a positive, intelligent relationship between business and government. During the past three decades, many automotive enthusiasts stood idly by and watched the onslaught of consumer advocates take on a recalcitrant American automobile industry with the help of the federal government. This battle, one that pitted those who disliked or even hated cars against those who all but forgot how to build them, brought about vehicles only marginally safer in low-speed crashes. It also created an antipathy between business and government that exists to this day. With guarded optimism, things have improved to some degree. Automobile manufacturers finally seem to have figured out that "safety sells." But safety will sell only to a point. Passive safety alone will not sustain better sales figures for cars and trucks. Companies like Volvo have had to face this fact and modify their cars and sales campaigns to meet customers' desires for performance and styling.

Recognizing the correct blend of styling, performance and passive safety is the key if the American automobile industry wants to build vehicles that are clearly superior to the foreign competition, win back their lost market share and keep Americans working.

When "Engine" Charlie Wilson, the former President of GM stated, "What is good for General Motors is good for the country," he should have said, "What is good for America is good for the American automobile industry and vice versa." This nation's past, present and future prosperity is directly linked to a healthy *domestic* auto industry. If America is to remain a powerful economic force in the world, the building of cars and trucks by U.S. corporations must flourish. Car lovers must step forward to guide the American automobile industry, pointing the way to the innovations necessary, keeping an indispensable part of our national economy vital.

Beyond these safety improvements, there are two other problems that need to be addressed concerning the automobile's impact on the America of tomorrow. First, upgrading existing roads and expanding our highway network across the country is an obvious necessity, but there is very little room or political will to build more roads in our larger cities. For the long term, the positive role that alternative transportation could play in relieving congestion and gridlock has to be further explored. In the battle of the big city, the automobile will not win. It is simply not an efficient enough form of transportation in an area of high population density. For most metropolitan areas, however, the automobile will remain the best way to travel across town. Traffic congestion has to reach a very high level before it is faster to leave your home, walk, bike, bus or taxi to a subway terminal—ride the train—then walk, bike, bus or taxi to your destination. The car is uniquely suited for 90 percent of America's surface transportation needs. Presently though, billions of gallons of gas and millions of hours are wasted because of automobile congestion in big cities across the country. This translates into smog and lost productivity every day of the year. Some short-term improvements can come from staggering work hours, expanding bus service and "micro" car pooling to reduce congestion for special events. Long term, new forms of alternative, nonautomotive transportation have to be pursued. One possibility is a mainline network of fast-moving trains branching outward to park-and-ride stations around the centers of our most densely populated cities. Already, there are many lightly used or abandoned railroad spurs whose right-of-way could be used for this new transportation system. It is of paramount importance that this new rail system interface well with our present and future roadways. Designing these rail networks to travel over or under existing roads guarantees the highest degree of safety and transportation efficiency. The possibility presents itself of living well out into the country, commuting a half-hour by freeway at 80 mph to a park-and-ride train stop for another half-hour trip into the city at the same speed. True country living could coexist with big city convenience if we take advantage of what the automobile and alternative transportation can provide. Exploiting the virtues of both systems could be an important step in maximizing our use of the automobile while reducing its negative impact on the environment and economy. If properly designed, this union of road and rail could bring an unprecedented period of economic revitalization to certain big cities in America. However, monitoring the changes in population and business trends over the next two decades will be critical to learn what form of transportation will work the best. Will the further expansion of the Internet revitalize small towns? The transfer and sale of electronic information is not contingent on location. Industry might diversify into rural America, creating a shift in population that could ease traffic congestion in some areas without alternative transportation. Will we work to improve the quality of life in our big cities? Nonautomotive, alternative transportation would be of benefit over the long term. Whatever happens, the automobile will continue be *the* mode of travel for most Americans in the foreseeable future.

The other major problem is the need to lower the automobile's impact on the environment. As faster driving becomes more prevalent, higher speeds mean higher engine rpms and, therefore, increased fuel use and more pollution. While the relationship of engine size, aerodynamics and miles per gallon is more complex than this suggests, it is, nonetheless, generally true. The faster you

Flexliner commuter train on a test run in Minneapolis, Minnesota, July, 1997. Using existing freight-rail lines, such trains could provide long-term improvement to transportation efficiency in heavily populated cities, especially when combined with easy park-and-ride access. Photo by the author

go, the more you pollute. Short term, we can shift our emphasis away from emissions inspections for all vehicles to cleaning up the one-in-ten that is a "gross polluter." Long term, cleaner air can come from alternative transportation and some alternative fuels like natural gas and hydrogen, but the majority of auto pollution will still remain. However, the specter of smog-filled U.S. cities does not have to be a part of the 21st century, either. The car companies are working very hard to clean up emissions, from the thriftiest econobox to the thirstiest sport utility vehicle. The reason? It's good business. Customers are slowly becoming more environmentally aware and the auto industry is simply meeting the demand. Granted, this trend is probably happening far too slowly for the ardent environmentalist, and too quickly for the laissez-faire capitalist (which probably means it's right on track). But there is no doubt that striving for cleaner emissions will remain part of building a better car for as long as the internal combustion engine is part of its design. And that motor is here to stay for at least another thirty to forty years. No other alternative, from fuel-cell technology to improved batter-

Dodge Intrepid ESX2. Courtesy of DaimlerChrysler Corp.

EV-1. © General Motors Corp. Used with permission of GM Media Archives

The road to a cleaner future. The Dodge ESX2 diesel-electric hybrid and General Motors electric vehicle EV-1. Until battery technology dramatically improves, electric vehicles will remain small in number through the first quarter of the 21st century. More promising is hybrid technology, generating electricity from a small, low-emissions internal combustion engine to power an electric motor to drive the wheels.

ies for electric vehicles, can match the cost-effective performance related to emissions. Already, hybrid vehicles are on the road which generate electricity from a small gas or diesel engine that powers an electric motor to drive the wheels. This design has already doubled fuel mileage with very low pollution. Environmental purists who hope to see true zero emissions electric vehicles recharged from sunlight or hydroelectric energy will have to wait a long time for these expensive, new technologies to become cost-effective and commonplace. Taking into account a global economy based on oil (*and* trillions of dollars in future oil profits at stake), it is clear where the ever cleaner energy for the next half century is coming from.

This brings us to the real bottom line: How to pay for all this Fast-and-Safe technology? How do we generate the billions of dollars necessary to fund these improvements when we are more than $5 trillion in debt?

While you may gripe about the annual amount of money the Internal Revenue Service removes from your paycheck, the truth is, we are all *under* taxed when it comes to the price of a gallon of gasoline. If you consider that Germany's price per gallon is nearly $4 and the rest of the industrialized nations of the world pay anywhere from $3 to $5 per gallon, we have allowed our price to remain too low for too long. Whether you like it or not, the days of cheap gas are quickly drawing to a close. There are too many transportation-safety and infrastructure problems that must be dealt with, and paying up front is the *only* way out. Funding higher-speed freeways and expanded alternative transportation systems will require, *at a minimum*, raising the gas tax ten cents per year for the next ten years. The average price of gas must reach the $2 mark by the end of the next decade if we are to bring about the changes needed to renovate and reinvent our transportation network. Here is where the desire for improvement runs headlong into the reluctance to pay. And the temptation to remain mired in the status quo of a polluted, gridlocked, modest-speed highway system with 40,000 fatalities per year pulls stronger on the pocketbook than the prospect of a clean, efficient, 100 mph transportation marvel with 20,000 annual fatalities—one that you rolled up your sleeves, worked hard and paid for. If we have the foresight to make this investment, it will pay back billions in dividends through reduced congestion, pollution, death and suffering, not to mention lower medical and auto insurance premiums—even though we would be driving faster than ever before. Paying up front also makes alternative transportation and the corresponding infrastructure more competitive. For truckers and other heavy fuel users, a system of tax credits and a more modest tax increase for diesel fuel should be mandated to ease the transition to higher operating costs. Further, phasing in higher gas taxes at two and a half cents per quarter would make the transition a bit easier to bear. If automobile manufacturers up their ante and support the program, this gas tax increase will do more to improve fuel economy than any federal mandate. This, and higher speeds, will move American cars and trucks to the vanguard of performance and safety design innovation. That translates to increased economic prosperity, making us more competitive at home and on the world market. Putting this Fast-and-Safe philosophy into action has one, final benefit that most national programs do not have: You get to see your money at work. Driving down *this* road to tomorrow means better pay, clearer skies and higher numbers on the speedometer—as well as fewer deaths behind the wheel.

Which brings us back to those of us who love cars. Now that all is said and done, and that little voice in the back of your mind whispers: "Gee, wouldn't it be great, but it'll never happen." That thought must be quelled. No great accomplishment, no positive accomplishment at all, ever comes from pessimistic thinking. One first has to get beyond the notion that it can't be done, and then strive to do it. The truth is, a better tomorrow doesn't just magically arrive one day. It starts *now* with inspiration, hard work and the willingness to pay up front to make it happen. Even so, a controversial philosophy like Fast and Safe requires a conservative approach to build the consensus necessary to implement it on the road. People's lives are at stake and the exhilaration of driving faster must be tempered with a responsible attitude that ensures greater safety. There is nothing wrong with driving your car at its limits, if the proper precautions have been taken on the right kind of road.

America needs to tackle some difficult problems to restore faith in itself and its ability to accomplish great things. There is no better place to start than on the road to tomorrow. As the new century and, in turn, the new millennium unfold, we must work together to leave a legacy of accomplishment that is the envy of future generations. By facing tough challenges head-on with unity, determination and honesty, we can *move* this great nation into an exciting future of our own making. The millions of us who love both car and country (and not necessarily in that order) have the responsibility to lead, inspire and mobilize the rest of the nation in this endeavor. Imagine what a hundred thousand or a million well-organized automotive enthusiasts could do: We would be invincible. So turning the philosophy of this book from the printed page into a reality on the road is the challenge for all of us who admire the subtle curve on a fender, marvel at the turn of an engine and thrill to the crest of a hill and the freedom of the open road beyond.

finis

STATISTICS

Death Rate per 100 Million Vehicle-Miles Traveled on the German *Autobahn**
and American Interstate Freeway System

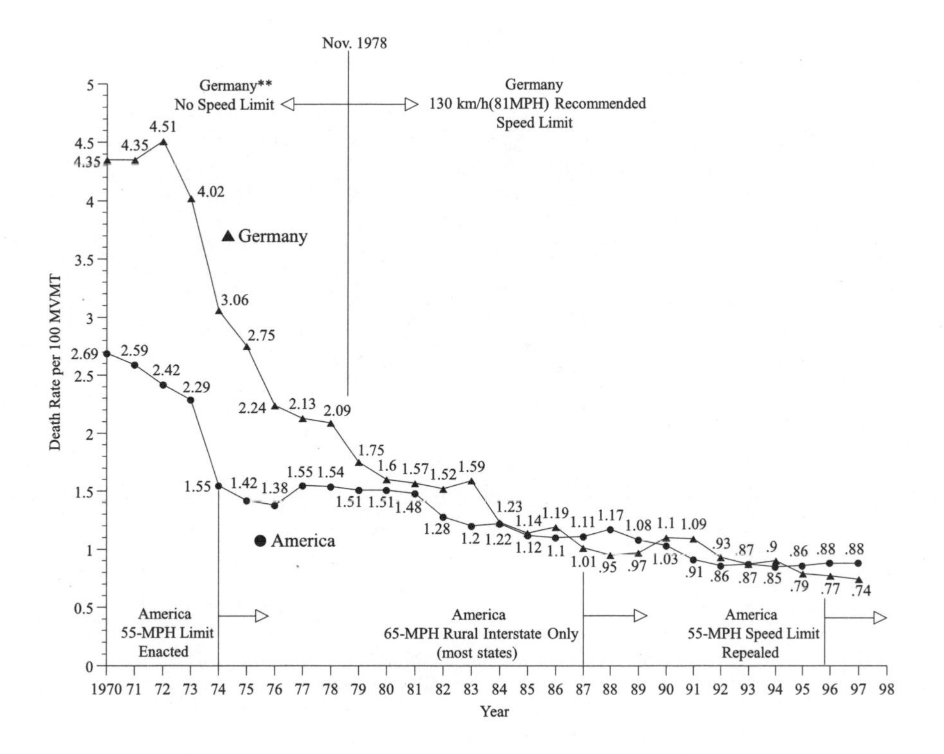

Note: **November 1973 to March 1974, 100km/h(63mph) speed limit on the *Autobahn*.

Sources:

Germany: Bundesminister Für Verkehr
America: Federal Highway Administration

*All figures are for the former West Germany

Travel Speeds on the American Rural Interstate Freeway and German *Autobahn**

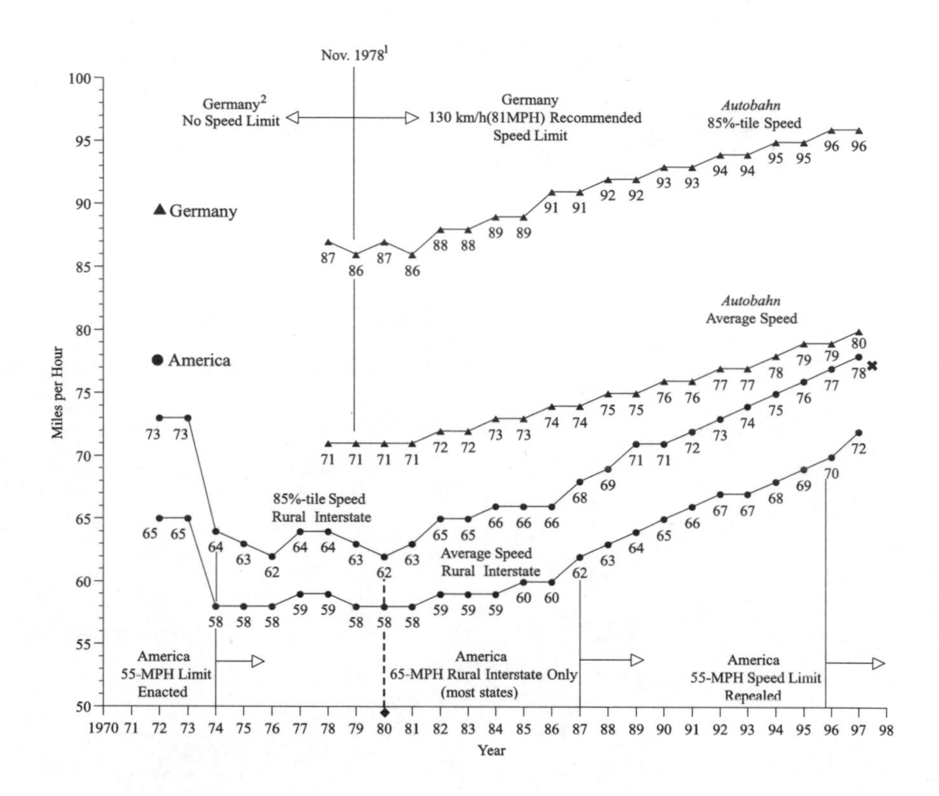

Notes: [1]Speed data before 1978 is unavailable for the *Autobahn.*
[2]November 1973 to March 1974, 100km/h(63mph) speed limit on the *Autobahn.*

◆1980-95 U.S. speeds represent speeds of all vehicles including congested roads, on hills and curves.
Free speeds likely 2-4mph higher.
1978-97 German *Autobahn* speeds are for free-moving vehicles.
✖Present-day, free-moving vehicle speeds can vary, plus or minus, 2-4 mph from eastern to western U.S.

Sources:
Germany: Bundesanstalt flr Strassenwesen
America: Federal Highway Administration

*All figures are for the former West Germany

Deaths by Year on the American Interstate System and the German *Autobahn**

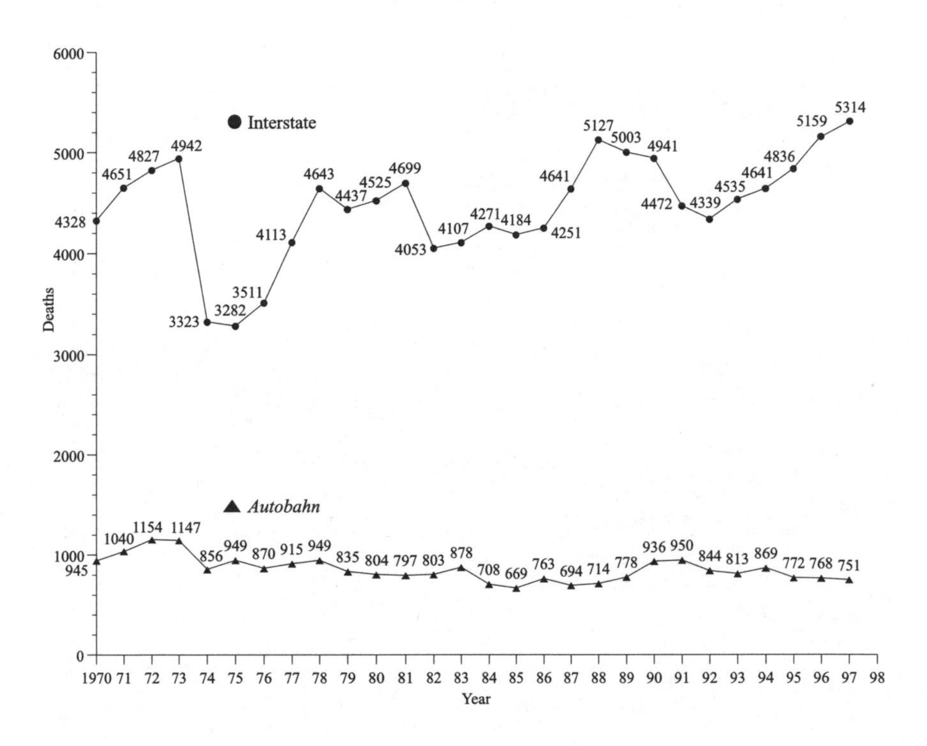

Sources:

Germany: Bundesminister Für Verkehr
America: Federal Highway Administration

*All figures are for the former West Germany

Deaths by Year for All Roads in America and Germany*

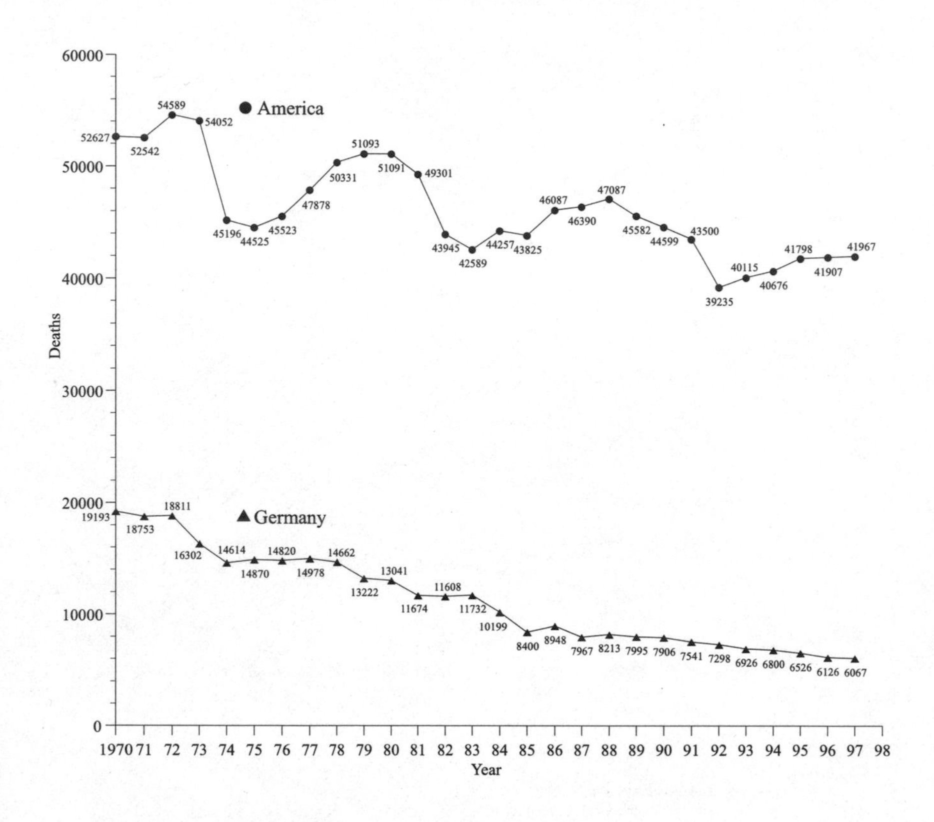

Sources:

Germany: Bundesminister Für Verkehr
America: Federal Highway Administration

*All figures are for the former West Germany

Death Rate per 100 Million Vehicle-Miles Traveled for All Roads in America and Germany*

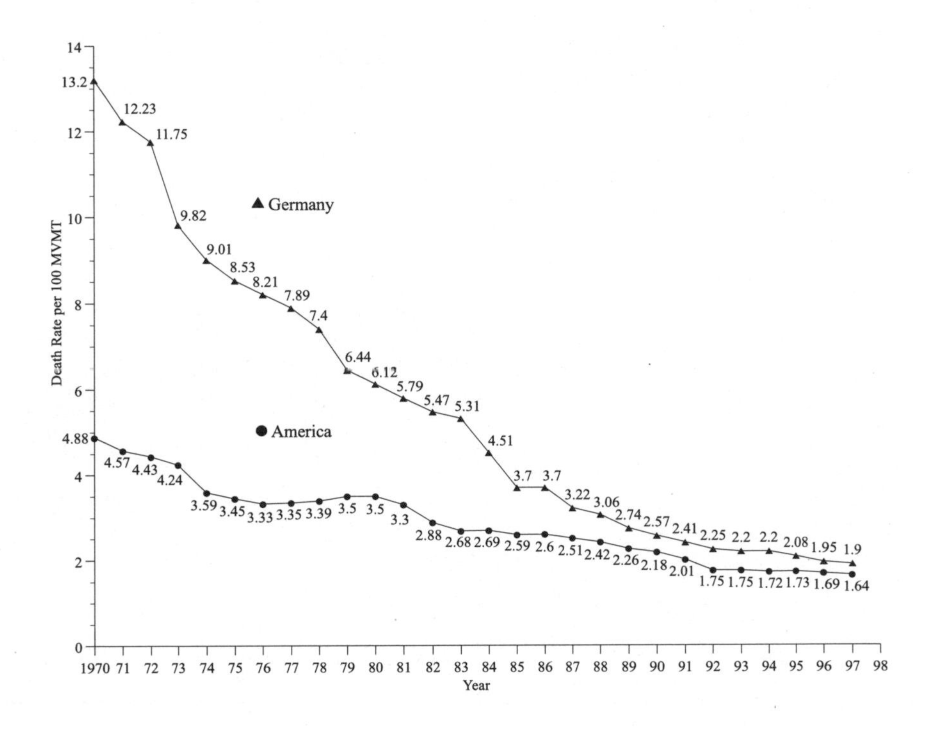

Sources:

Germany: Bundesminister Für Verkehr
America: Federal Highway Administration

*All figures are for the former West Germany

Motor-Vehicle Death Rate per 100,000 Population for America and Germany*

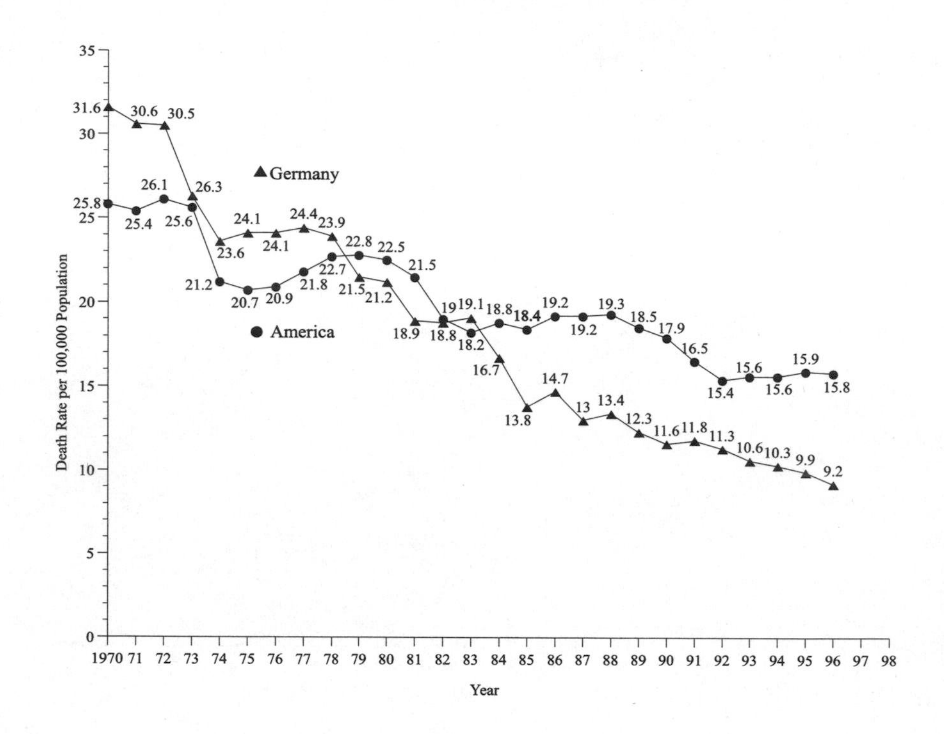

Sources:

Germany: Bundesminister Für Verkehr
America: Federal Highway Administration

*All figures are for the former West Germany

Motor Vehicle-Miles Traveled, Traffic Fatalities and Death Rate per 100 million VMT for America (1940-1996)

Year	Miles Driven(Billions)	Fatalaties	Death Rate
1940	302	32914	10.9
1941	334	38142	11.4
1942	268	27007	10.1
1943	208	22727	10.9
1944	213	23165	10.9
1945	250	26785	10.7
1946	341	31874	9.4
1947	371	31193	8.4
1948	398	30775	7.7
1949	424	30246	7.1
1950	458	33186	7.2
1951	491	35309	7.2
1952	514	36088	7
1953	544	36190	6.6
1954	561	33890	6
1955	603	36688	6.1
1956	628	37965	6
1957	647	36932	5.7
1958	665	35331	5.3
1959	700	36223	5.2
1960	719	36399	5.1
1961	737	36285	4.9
1962	767	38980	5.1
1963	805	41723	5.2
1964	846	45645	5.4
1965	888	47089	5.3
1966	926	50894	5.5
1967	964	50724	5.3
1968	1016	52725	5.2
1969	1062	53543	5
1970	1110	52627	4.9
1971	1179	52542	4.6
1972	1260	54589	4.4
1973	1313	54052	4.2
1974	1281	45196	3.6
1975	1328	44525	3.5
1976	1402	45523	3.3
1977	1467	47878	3.4
1978	1545	50331	3.4
1979	1529	51093	3.5
1980	1527	51091	3.5
1981	1553	49301	3.3
1982	1595	43945	2.9
1983	1653	42589	2.7
1984	1720	44257	2.7
1985	1774	43825	2.6
1986	1835	46087	2.6
1987	1921	46390	2.5
1988	2026	47087	2.4
1989	2096	45582	2.3
1990	2148	44599	2.2
1991	2172	41508	2
1992	2247	39250	1.8
1993	2297	40150	1.8
1994	2358	40716	1.7
1995	2423	41817	1.7
1996	2482	41907	1.7

Source: Federal Highway Administration

ACKNOWLEDGMENTS

It must be duly noted that many individuals and organizations generously donated time and expertise while not necessarily agreeing with every aspect of this book or knowing its full content.

Allgemeiner Deutscher Automobil-Club (ADAC), Munich, Germany. Christine Baker, Photographic Specialist, Pennsylvania Turnpike Commission. Captain Steve Barry, Montana Highway Patrol. James J. Baxter, President, National Motorists Association, BMW of North America. Der Bundesminister für Verkehr, Matthias Wissmann und seine Vorgänger, Bonn, Germany. Robert S. Chirinko, Emory University. Julie Anna Cirillo, U.S. Department of Transportation (DOT) - Federal Highway Administration (FHwA). DaimlerChrysler Corp. Clarence Ditlow III, Executive Director, Center for Auto Safety. Joanne Domka, GM Media Archives. Sergeant James M. Eagan, New York State Police, (Ret.). Ford Motor Company. Alex Gabbard. Duke Ganote, National Motorists Association, OH. Laura Giammarco, Time-Life Syndication. Dr. Paul Gifford, Ovonic Battery Company, Troy, MI. Frank Goddard, Administrative Assistant to Kansas Highway Patrol, Kansas Turnpike Authority. James M. Rashid, B.F. Goodrich Corp. Helgard Gries-Rittinghaus, Bundesanstalt für Straßenwesen (BASt). Nevin Harwick. Jay Hawkinson for chapter symbols. Kim Hazelbaker, Vice President, Highway Loss Data Institute. Janet M. Healy, Manager of Facilities, Liberty Mutual Insurance Research Center. George E. Hoffer, Virginia Commonwealth University. Susan A. Hollis, Electric Auto Association. Herbert Horak Fahrschule, Pfarrkirchen, Germany. The Father of the 55 mph speed limit, the late James J. Howard, U.S. Representative, NJ. Robert Hurst, Tribune Media Services. Patricia Isaacs, Parrot Graphics, for maps, cross section drawings & road signs. Wolfgang Kuchenbauer für die Münchener Fahrt herum. John Lamm, Editor at Large, *Road & Track* magazine. Ben Langer, U.S. DOT-National Highway Traffic Safety Administration (NHTSA). Chet Lasell, Director of Public Relations, Liberty Mutual Insurance. Colonel D. Roger Ledding, Minnesota State Patrol, (Ret.). Don Leeper, Stanton Publication Services, Inc., St. Paul, MN. Axel "Open Road" Madsen. Maryland Department of Transportation. Sergeant Mike McBride, Minnesota State Patrol, (Ret.). Lieutenant Monty McCord. Mercedes-Benz AG, Stuttgart, Germany. Susan Nelson for rights and permissions assistance. Olive Nerem, MN/DOT-Library. Jerry Nowlin, National Motorists Association, Maryland. Gary Olson for making me look presentable. Brian O'Neill, President, Insurance Institute for Highway Safety. George Parker, Associate Administrator for Research and Development, U.S. DOT-NHTSA, (Ret.). Lieutenant Mark Peterson, Minnesota State Patrol. Porsche AG, Weissach, Germany. Bruce Priebe, Twin Cities Section, Mercedes-Benz Club of America. Carolyn A. Rask for her great tolerance of my dreams. Douglas W. Rask as an entrepreneurial example. Julia M. Rask for unflagging support. Kathi Redding for computer literacy. Cheri Register and Amy Lindgren for giving my manuscript more coherent sentence structure. Curt "Drive to Survive" Rich, Star Motor Cars, Houston, TX. Henry Sandhusen, Chief, Traffic Control Programs Team, U.S. DOT-NHTSA. Corporal Ed Sanow. David M. Seiler, Police Traffic Services, U.S. DOT-NHTSA. Randall T. Skaar, Patterson & Keough, P.A. Dieter Slezak, GraF/X, Mpls., MN. Leo und Gertrud Slezak, Pfarrkirchen, Germany. Eric Smith, Capital-Gazette Newspapers, Inc. Todd Spencer, Owner-Operator Independent Drivers Association. Mark P. Stehrenberger for making my vision of the future come to life. Guy Spangenberg, Spangenberg Photography, Orange, CA. Charles F. Timberg. Sergeant Larry Tolar, Colorado State Patrol. Andrea Twiss-Brooks, John Crerar Library, University of Chicago. Dennis Unsworth, Montana Department of Transportation. Mike Valente. Volvo Cars of North America. Louis C. Wendling, Twin Cities Section, Mercedes-Benz Club of America. Dave Wiese for help with charts & graphs. Benjamin Worel & Duane S. Young, MN/ROAD, MN/DOT. And a final thanks to those who wished not to be named in these ranks.

BIBLIOGRAPHY

Books

American Association of State Highway and Transportation Officials (AASHTO). *The States and the Interstates.* Washington, D.C.: AASHTO, 1991.

Bel Geddes, Norman. *Magic Motorways.* New York City: Random House, 1940.

Bondurant, Bob (with James Blakemore). *Bob Bondurant on High Performance Driving.* Osceola, WI: Motorbooks International Publishers, 1982.

Bundesministerium für Verkehr (BMV). *Roads in Germany.* Bonn, Germany: BMV, 1994.

Eagan, James M. *A Speeder's Guide to Avoiding Tickets.* Newport, NY: Caretaker Publishing, 1990.

Eastman, Joel. *Styling vs Safety: The American Automobile Industry and the Development of Automotive Safety 1900–1966.* Lanham, MD: University Press of America, 1984.

Egan, Philip. *Design and Destiny: The Making of the Tucker Automobile.* Orange, CA: On the Mark Publications, 1989.

Federal Highway Administration (FHwA). *Report on the 1992 U.S. Tour of European Concrete Highways.* Washington, D.C.: FHwA, 1992.

Finch, Christopher. *Highways to Heaven.* New York City: HarperCollins, 1992.

Gabbard, Alex and Squire Gabbard. *Fast Muscle: America's Fastest Muscle Cars.* Lenoir City, TN: Gabbard Publications, 1990.

Haddon, William. *Accident Research: Methods and Approaches.* New York City: Harper and Row, 1964.

Halberstam, David. *The Reckoning.* New York City: William Morrow & Co., 1986.

Hazleton, Lesley. *Confessions of a Fast Woman.* New York City: Addison-Wesley, 1992.

Iacocca, Lee (with William Novak). *Iacocca: An Autobiography.* New York City: Bantam, 1984.

Jerome, John. *The Death of the Automobile.* New York City: W.W. Norton, 1972.

Kay, Jane Holtz. *Asphalt Nation.* New York City: Crown Publishers, 1997.

Kirschbaum Verlag. *Autobahnen in Deutschland.* Bonn, Germany: Kirschbaum, 1979.

Madsen, Alex. *Open Road: Truckin' on the Biting Edge.* New York City: Harcourt Brace Jovanovich, 1982.

Mandel, Leon. *Driven: The American Four-Wheeled Love Affair.* New York City: Stein and Day, 1977.

Marsh, Peter and Peter Collett. *Driving Passion: The Psychology of the Car.* Boston, MA: Faber and Faber, Inc., 1987.

McCarry, Charles. *Citizen Nader.* New York City: Saturday Review Press, 1972.

McCord, Monty. *Police Cars: A Photographic History.* Iola, WI: Krause Publications, 1991.

Menken, Eugen. *Der 7. Sinn: Der Grosse Ratgeber zur Erfolgreichen ARD-Fernsehserie.* Cologne, Germany: Naumann und Göbel, 1986.

Nader, Ralph. *Unsafe at Any Speed.* New York City: Grossman, 1965.

Niemann, Harry. *Béla Barényi, The Father of Passive Safety.* Stuttgart, Germany: Mercedes-Benz AG, 1995.

Patton, Phil. *Open Road: A Celebration of the American Highway.* New York City: Simon and Schuster, 1986.

Pettifer, Julian and Nigel Turner. *Automania: Man and the Motor Car.* Boston, MA: Little, Brown, 1985.

Rich, Curt. *Drive to Survive.* Osceola, WI: Motorbooks International Publishers, 1998.

Rose, Mark. *Interstate: Express Highway Politics, 1941–1956.* Lawrence, Kansas: The Regents Press of Kansas, 1979.

Ross, Laurence H. *Confronting Drunk Driving: Social Policy for Saving Lives.* New Haven, CT: Yale University Press, 1992.

Shank, William H. *Vanderbilt's Folly: A History of the Pennsylvania Turnpike.* York, PA: American Canal and Transportation Center, 1964.

Shnayerson, Michael. *The Car that Could: The Inside Story of GM's Revolutionary Electric Vehicle.* New York City: Random House, 1996.

Tomerlin, John and Dru Whitledge. *The Safe Motorist's Guide to Speed Traps.* Chicago: Bonus Books, Inc., 1991.

Turner, Richard and David Shelton. *Accident Avoidance and Skid Control.* Hutchins, TX: NAPD Publishing Company, 1979.

Wallis, Michael. *Route 66: The Mother Road.* New York City: St. Martin's Press, 1990.

Whitmore, John. *Superdriver: Exercises for High-Concentration, High-Performance Driving.* Osceola, WI: Motorbooks International, 1988.

Yates, Brock. *The Decline and Fall of the American Automobile Industry.* New York City: Empire Books, 1983.

Federal Government Hearings

Haddon, William, Jr. *Remarks Concerning Structures and Materials.* Presented at the Conference on Basic Research Directions for Advanced Automotive Technology. U.S. Department of Transportation Systems Center. Boston, MA, February 13, 1979.

Oliver, David. Office of Chief General Counsel, Federal Highway Administration. *State of Arizona 55-Mile-Per-Hour National Maximum Speed Limit Compliance Hearing.* U.S. Department of Transportation, Federal Highway Administration. Phoenix, AZ, March 13, 1985.

——. *Maryland 55 MPH Limit Hearing.* U.S. Department of Transportation, Federal Highway Administration. Baltimore, MD, May 22, 1985.

——. *State of Vermont 55-Mile-Per-Hour Noncompliance Hearing.* U.S. Department of Transportation, Federal Highway Administration. Montpelier, VT, March 26, 1985.

O'Neill, Brian. *Statement to Hearings on the 55 MPH National Maximum Speed Limit.* Presented before the House of Representatives Committee on Public Works and Transportation, Subcommittee on Surface Transportation. Washington, D.C., March 18, 1987.

Subcommittee on Surface Transportation, U.S. House of Representatives. *Hearing to Examine the Enforcement and Monitoring of the 55 MPH Speed Limit.* Washington, D.C., July 23, 1985.

Pamphlets

Cupper, Dan. *The Pennsylvania Turnpike: A History.* Lebanon, PA: Applied Arts Publishers, 1990.

Kansas Turnpike Authority. *1995 Forty-Third Annual Report.* 1995.

National Motorists Association. *Motorists' Guide to State Traffic Laws.* Dane, WI: NMA, 1992.

Periodicals

Aukofer, Matthew. "Dyno Smogged." *Autoweek,* November 23, 1998.

Barry, David. "Hammerin' Down the Road." *Autoweek,* June 23, 1986.

Bedard, Patrick. "The 55 mph Speed Limit." *Car and Driver,* July, 1983.

——. "What's the Deal on Electric Cars." *Car and Driver,* May, 1992.

——. "What if Airbags Don't Work on Smart People." *Car and Driver,* November, 1994.

——. "Auto Insurance: State Farm Caught Faking Facts." *Car and Driver,* January, 1995.

——. "Still Smoggy After All These Years." *Car and Driver,* April, 1995.

——. "One Drink Over the Line." *Car and Driver,* September, 1998.

Greenwald, John. "Can GM Survive in Today's World." *Time,* November 9, 1992.

Griffen, Larry. "The State of the Interstates." *Car and Driver,* August, 1986.

Insurance Institute for Highway Safety. *Status Report.* Monthly Issues. 1989–1999.

Jennings, Gordon. "The Shuck and Jive of 55." *Motor Trend,* November, 1984.

Kacher, Georg. "German Joy Ride." *Car and Driver,* June, 1986.

Koepp, Stephen. "Gridlock!" *Time,* September 12, 1988.

Lang, John S. "The Automobile Turns 100." *U.S. News and World Report,* September 30, 1985.

Lankard, Tom. "Fifty-Five Under Fire." *Autoweek,* November 24, 1986.

——. "Truck Law Gets Semi Tough." *Autoweek,* November 24, 1986.

Martz, Larry. "Does Speed Kill?" *Newsweek,* July 21, 1986.

McGinn, Daniel and Adam Rogers. "Operation: Supercar." *Newsweek,* November 23, 1998.

Moynihan, Daniel Patrick. "Epidemic on the Highways." *Reporter,* April 30, 1959.

Nader, Ralph. "The Safe Car You Can't Buy." *The Nation,* April 11, 1959.
National Motorists Association. *NMA News*. Bi-monthly issues, 1982–1999.
O'Donnell, Jayne. "Are Teens Unteachable?" *Autoweek,* April 21, 1997.
Overend, Robert. "55 mph: Do We Still Need It?" *Traffic Safety,* Vol. 85, No. 4, July/ August, 1985.
Sielski, Matthew. "What Should the Maximum Speed Limit Be?" *Traffic Engineering,* September, 1956.
Simanaitis, Dennis. "Electric Vehicles." *Road & Track,* May, 1992.
——. "Let's Get Virtual." *Road & Track,* July, 1994.
——. "ABS: Putting a Stop to it All." *Road & Track,* July, 1997.
Taylor, Nick. "Roads That Bind Us, Interstate Highways: A Guide to Rediscovering America." *Travel Holiday,* August, 1990.
Thomas, Bill. "Interstate Highway System: Providing the Shortest Distance Between Two Points." *Holiday Inn,* January, 1969.

Technical and Research Reports

American Association of State Highway and Transportation Officials (AASHTO). *A Summary of Surface Transportation Investment Requirements 1988–2020.* Washington, D.C.: AASHTO, 1988.
——. *Highway Safety Strategic Plan 1991–2000.* Washington, D.C.: AASHTO, 1990.
——. *New Transportation Concepts for a New Century.* Washington, D.C.: AASHTO, 1989.
——. *Transportation Safety in 2020.* Washington, D.C.: AASHTO, 1989.
Bastarache, Gerald. *Beyond Gridlock.* Washington, D.C.: Highway Users Federation, 1988.
Bundesanstalt für Strassenwesen. *Periodische Analyse des Verkehrs-Ablaufs im Autobahnnetz (Entwicklung des Geschwindigkeits- und Abstandverhaltens).* 1991.
Bundesministerium für Verkehr (BMV). *Unfallverhütungsbericht Strassenverkehr.* Biyearly Issues. 1985–1995.
——. *Geschwindigkeit und Verkehrssicherheit im Straßenverkehr.* 1997.
——. *Sicherheit im Strassenverkehr.* 1996.
——. *Verkehr in Zahlen* 1997. 1997.
Center for Auto Safety. *Active Seat Belts.* 1985.
——. *Air Bags and Passive Seat Belts.* 1985.
——. *Design Standards and Highway Resurfacing Restoration and Rehabilitation.* 1977–1981.
——. *Fuel Economy Standards.* 1977–1980.
——. *Highway Construction Zone Safety and Traffic Systems Compatibility.* 1977–1980.
——. *Highway Safety Improvement Program.* 1978–1981.
——. *Interior Protrusions in Motor Vehicles.* 1980.
——. *Need for Advanced Federal Safety Standards.* 1980.
——. *Pedestrian Safety and Exterior Protrusions.* 1980.

——. *Post-1985 Fuel Economy Standards.* 1985.

——. *Public Participation in Highway Decision Making.* 1975–1980.

——. *Washington Under the Influence.* 1976.

——. *The Yellow Book Road: The Failure of America's Roadside Safety Program.* 1974.

Chirinko, Robert S. and Edward P. Harper, Jr. "Buckle Up or Slow Down? New Estimates of Offsetting Behavior and Their Implications for Automobile Safety Regulation." *Journal of Policy Analysis and Management.* Vol. 12, No. 2, pp. 270–296, 1993.

Cirillo, Julie. "Interstate System Accident Research—Study II—Interim Report II." *Public Roads.* Vol. 35, No. 3. 1968.

Crandell, Frank. *Packaging the Passenger—Design for Collision Project, Survival Car II.* ASME Publication, 1962.

Dawson, Harry S., Jr. "Analysis of Fatal Accident Trends on Maryland Highways, 1970–1976." *Public Roads.* September, 1979.

Fatal and Injury Accident Rates on Public Roads in the United States. Annual Issues. U.S. Department of Transportation, Federal Highway Administration. 1968–1997.

Federal Highway Administration. *Effects of Raising and Lowering Speed Limits.* 1992.

Hauer, Ezra. "Accidents, Overtaking and Speed Control." *Accident Analysis and Prevention.* Vol. 3, pp. 1–13. Elmsford, New York: Pergamon Press, Inc., 1971.

Hoffer, George E. and Steven P. Peterson. "The Impact of Airbag Adoption on Relative Personal Injury and Absolute Collision Insurance Claims." *Journal of Consumer Research,* Inc. Vol. 20. March, 1994.

Insurance Institute for Highway Safety. *Facts.* Annual Issues. 1991–1996.

Johnston, Edward J. "Slow Traffic Laws Waste Fast Roads." *Traffic Engineering.* July, 1956.

Joscelyn, Kent B. *Maximum Speed Limits—A Study for the Selection of Maximum Speed Limits.* Vol. 1, NTIS No. PB-197 373, U.S. Department of Transportation, National Highway Traffic Safety Administration. October, 1970.

——. *Maximum Speed Limits—The Development of Speed Limits: A Review of the Literature.* Vol. 2, NTIS No. PB-197 374, U.S. DOT, NHTSA. October, 1970.

Liberty Mutual Insurance. *The Cornell-Liberty Safety Car.* 1957.

Maryland State Police, Traffic Programs Planning Unit. *Discussion Paper on Enforcement of the 55 NMSL: A Historical Perspective.* 1984.

McBride, Michael S. *ABS Brakes, My Experiences.* Unpublished Research Report. 1994.

Mercedes-Benz. *30 Years of Accident Testing, 50 Years of Passive Safety.* 1989.

Minnesota Department of Transportation. *Minnesota 85%-tile Speed History.* 1995.

Motor Vehicle Manufacturers Association. *Facts and Figures '91.* 1991.

National Research Council. *55: A Decade of Experience.* (Special Report 204). Washington, D.C.: Transportation Research Board, 1984.

National Safety Council. *Accident Facts.* Annual Issues. 1970–1998.

——. *Speed Regulation.* Committee on Speed Report. 1941.

Nevada Department of Transportation. *Annual Speed Monitoring Reports.* 1996–1997.

Northwestern University Traffic Institute. *Speed Offenses.* 1988.

——. *Techniques for Radar Speed Detection.* 1988.

Peltzman, Sam. "The Effects of Automobile Safety Regulation." *Journal of Political Economy.* Vol. 83, pp. 677–725. August, 1975.

Solomon, David. *Accidents on Main Rural Highways Related to Speed, Driver and Vehicle.* U.S. DOT, FHwA, 1964.

Sumner, Roy. *A Variable Speed Limit System for Freeways.* ITE Compendium of Technical Papers, 1988.

Sumner, Roy and C.M. Andrews. *Variable Speed Limit System.* FHWA No. RD-89-001. U.S. DOT, FHwA, 1990.

Tignor, Samual and Davey Warren. *Driver Speed Behavior on U.S. Streets and Highways.* ITE Compendium of Technical Papers, 1990.

Verband der Automobil Industrie. *Fakten Gegen Tempolimit Auf Autobahnen.* 1985.

West, Leonard B., Jr. and J.W. Dunn. "Accidents, Speed Deviation and Speed Limits." *Traffic Engineering.* July, 1971.

INDEX